SpringerBriefs in Applied Sciences and Technology

W0234522

SpringerBriefs present concise summaries of cutting-edge research and practical applications across a wide spectrum of fields. Featuring compact volumes of 50 to 125 pages, the series covers a range of content from professional to academic.

Typical publications can be:

- A timely report of state-of-the art methods
- An introduction to or a manual for the application of mathematical or computer techniques
- A bridge between new research results, as published in journal articles
- A snapshot of a hot or emerging topic
- An in-depth case study
- A presentation of core concepts that students must understand in order to make independent contributions

SpringerBriefs are characterized by fast, global electronic dissemination, standard publishing contracts, standardized manuscript preparation and formatting guidelines, and expedited production schedules.

On the one hand, **SpringerBriefs in Applied Sciences and Technology** are devoted to the publication of fundamentals and applications within the different classical engineering disciplines as well as in interdisciplinary fields that recently emerged between these areas. On the other hand, as the boundary separating fundamental research and applied technology is more and more dissolving, this series is particularly open to trans-disciplinary topics between fundamental science and engineering.

Indexed by EI-Compendex, SCOPUS and Springerlink.

Habeeb Adewale Ajimotokan

Research Techniques

Qualitative, Quantitative and Mixed Methods Approaches for Engineers

 Springer

Habeeb Adewale Ajimotokan
Department of Mechanical Engineering
University of Ilorin
Ilorin, Nigeria

ISSN 2191-530X   ISSN 2191-5318 (electronic)
SpringerBriefs in Applied Sciences and Technology
ISBN 978-3-031-13108-0   ISBN 978-3-031-13109-7 (eBook)
https://doi.org/10.1007/978-3-031-13109-7

This Springer imprint is published by the registered company Springer Nature Switzerland AG
The registered company address is: Gewerbestrasse 11, 6330 Cham, Switzerland

This book is dedicated to the Supreme Being, the Almighty for sparing my life up to this day and seeing me through every moment of my life, and also to my wife and children for their enduring patience and support throughout the writing period.

Preface and Acknowledgements

This book, a maiden edition, is packed with several years of classroom teaching and research activities of a seasoned tutor of research techniques in engineering. Its goals are to assist in dealing with the required generic and soft skills to successfully conduct research and technology development by undergraduate and graduate engineering students or for a successful research and technology development career in academia and industry. Research in the engineering field requires a set of specific skills, distinctively unalike in approach to other scientific communities. The archival literature or bibliographic databases that might be beneficial for literature searching and review differ, and the challenge of choosing to publish the research findings as a patent, journal or conference paper is a dilemma most research technique books, typically do not consider.

Notwithstanding the engineering discipline or research topic, the tasks ahead of any researcher undertaking engineering research include the necessity to choose a researchable topic; state the problem statement; identify the aim and objectives of the study; plan and design an appropriate methodology; collect, collate, and analyse findings; and write a well-written report. Though engineering research projects often demand hi-tech, sophisticated techniques, and typically, computing and statistical knowledge. It is entirely possible to carry out a meaningful research project without the expertise in computing and statistical knowledge. Moreover, conducting a research project is best learnt by doing it; however, numerous efforts and goodwill might be wasted and dissipated due to lack of adequate planning and design of the project.

Engineering research, typically conducted in the academia or industries, are industry-driven and their deliverables are, thus bound by timelines and driven by customers' requirements, in particular, for research carried out in research or industry laboratories. Also, engineering research demands some cautious ways and manners of response to reviewers' criticisms of submitted conference or journal paper manuscripts, or criticisms of submitted project reports by project supervisors, examiners, or lab managers in academia or industry. So, regarding these, tips are provided.

In this book, the technique of concentrating on the research project by stating the research problem and phrasing research questions, the art of using libraries, creating an effective literature search strategy and review, developing a research plan, and writing a project report and its presentations, coupled with specialised or focused aspects, associated with a stream of engineering, such as the use of commercial software for modelling and simulations, and citation management, were all discussed.

I have been encouraged throughout the writing of the book chapters by the interest of former and present undergraduate and graduate students, colleagues, friends, and associates. I am, in particular, grateful to *Mr. Abdul Wahab Lawal*, an educationist, poet and editor who proofread and commented on the drafts of all the chapters, and other colleagues who have encouraged and provided insights during the writing and production processes.

Ilorin, Nigeria   Habeeb Adewale Ajimotokan

About This Book

This book entitled *Research Techniques: Qualitative, Quantitative and Mixed Methods Approaches for Engineers* provides a hands-on guide towards conducting state-of-the-art engineering research. It gives pragmatic, step-by-step instructions that covered every stage in engineering research, from choosing a topic through to the presentation of research outcomes. The topics covered include the introduction and basic concepts of engineering research; research problem and questions; use of libraries, literature search and review; developing a research plan; and project report writing and presentations. This book, written in particular for engineers across all engineering disciplines, is ideal for undergraduate and graduate students; first-time or beginner researchers and academics intending to launch their research studies and careers; and would be of service as a reference resource to researchers, and those in the academia and industry.

The unique features of this book are its provision of unique engineering perspectives; discussion of research planning and implementation tools for engineering researchers, provision of pragmatic, step-by-step instructions for conducting engineering research; presentation of generic and soft skills required for impact; and offer of outstanding service as a reference resource to researchers and those in the academia and industry.

Contents

About the Author

Habeeb Adewale Ajimotokan (Ph.D., R.Eng.) is an Associate Professor and former Postgraduate Programmes Coordinator at the Department of Mechanical Engineering, University of Ilorin, Ilorin, Nigeria. He has been teaching or co-teaching Research Techniques in Engineering for several years. He has authored or co-authored several scientific articles, which comprise refereed journal articles, edited conference papers, book chapters and a book in reputable outlets of international standard; reflecting his scholarly prowess, contributions to contemporary research, knowledge, and consultancy services.

Chapter 1
Introduction and Basic Concepts of Engineering Research

Abstract The objectives of this chapter are to

(i)   Define the term engineering research and specify the purpose for engineering research;
(ii) Outline the different forms and types of engineering research;
(iii) Define the term research methodology, and differentiate between research methodology and methods;
(iv) Define the terms research approach, design and method, and specify their basic types; and
(v)   Discuss the phases in conducting engineering research.

Keywords Research purpose · Research methodology · Research approach · Experimental research design · Non-experimental research design · Research method · Research process

1.1 Introduction

Engineering research is frequently executed to establish, sustain, advance and promote innovative knowledge based on the application of scientific principles for the benefit of humanity. Engaging in these, engineering research seeks to probe into the unknown or mystery, by seeking solutions to research problems or answers to research questions and applying them. Usually, the mere consideration of doing engineering research, for some, might be worrying, scary and frightening, which in no circumstance should be. Dedicated engineering students, scholars, or researchers, who are imaginative, creative, and innovative, and as well can persevere when given feedback, i.e., positive criticism, could successfully develop their *research plan*, execute their *research project*, write and defend their *project report*, and present and publish *conference* and *journal papers.*

Like any scientific research, the *engineering research process* typically begins with defining an in-depth specific area of research interest (research theme), involving a problem, question, or phenomenon, which a researcher or a team of researchers wishes to address, seeking solutions, answers, postulation of theories, generalisations or principles. But unlike scientific research, engineering research ends with the application of scientific solutions, postulation of theories, generalisations or principles. This engineering research process that begins with defining an in-depth specific area of research interest, involving a problem, question, or phenomenon, would culminate in choosing a research topic, which should be intelligible, answerable, and contribute to the knowledge gaps in the subject area. The specific area of research interest, for instance, may be so broad that the researcher cannot execute it as a single engineering research project. Thus, the researcher can scale down the research topic to a topic that can be convincingly studied as a research project within a specific time frame.

1.2   Engineering Research

The word '*research*' itself originated from the word '*recherchier*'—a French word meaning to search or to investigate thoroughly [1]. So, research can be expressed as the systematic application of scientific methods for solving problems, investigating materials and sources to establish facts and deducing new inferences. In other words, research is a process of inquiry, investigation, scrutiny, and discovery, probing and finding solutions to problems and answers to questions, and generating those which currently do not exist. Hence, *engineering research*, on the other hand, involves the systematic application of the principles of scientific research, based on scientific methods, wherein experiments (observations or simulations), theories, generalisations, and corollaries, among others, can be derived from an existing scientific body of knowledge and verified by experts [2], while the verification by experts, called *peer review*, is not foolproof or infallible but constitutes the best verification and research outcome validation process. Though engineering research is precisely executed based on the scientific method, it tends towards a systematic application of scientific principles for infrastructure, products, and service delivery.

There is no gainsaying that scientific and engineering research are not mutually exclusive. Engineering research involves the processes of investigating problems, questions, or concerns through the application of scientific methods and collecting data in a strictly controlled situation for application, prediction, and explanation. Numerous engineers, and hence engineering research, are carried out by applying the principles of scientific research for investigating concepts with the long-term objective of seeking solutions to the research problem or answers to research questions for the practical implementation of the associated research findings. Because there are no distinct distinctions between both fields, several conference proceedings and scientific publications cover both science- and engineering-based research. However, an engineering research project might appear tedious and challenging, but it can be made easy and exciting through a painstakingly careful organisation of the research project planning or design.

1.3 Research Purposes

The purposes of the research are to provide solutions to the research problems and or answers to research questions by employing scientific methods. Though each piece of research has its specific purpose, engineering research is predominantly executed for the purpose of [3, 4]

(i) *Exploration*—undertaken typically to investigate or gain insight into poorly understood phenomena, discover or identify vital parameters or variables, and generate problems or questions for further research;

(ii) *Description*—undertaken to document and characterise the phenomenon of interest;

(iii) *Clarification*—a process undertaken to explain the cause(s) of a phenomenon;

(iv) *Understanding*—undertaken to comprehend a process, system, interaction, and phenomenon, among others;

(v) *Prediction*—a process undertaken to forecast events, future outcomes, physics of behaviour, and patterns based on established phenomena;

(vi) *Diagnosis*—undertaken to determine the frequency of occurrence or association of something with the object in view;

(vii) *Experimentation*—undertaken to conduct scientific procedures, in particular, in the laboratory so as to test, support, refute or validate the hypothesis; and

(viii) *Evaluation*—*a process* undertaken to make judgement(s) about quality of objects or events or assigning values to them.

1.4 Forms of Research

The primary *forms of research* are scientific research, research in the humanities, historical research, and artistic research [5]. S*cientific research* involves the systematic means of data collection and exploiting innovation to provide scientific information, generalisations, principles or theories for the explanation of nature and its properties. Scientific research makes practical applications possible and is mostly funded by government agencies, non-profit or business research foundations, and by charity and private groups [6], which include several companies. The *research in the humanities* involves diverse methods like semiotics and a distinctive, more relativist epistemology. Scholars in the humanities typically do not search for the ultimate, brief solutions to a problem or answers to questions, but rather examine the challenges and details encompassing it. The context, which might be cultural, social, ethnic, political, or historical, is always important. H*istorical research*, embodied in the historical method, is a good example of research in the humanities. A*rtistic research*, also called practice-based research, takes the form in which creative works are contemplated for the research and object of research itself. Unlike scientific research, artistic research is a debatable body of thought that provides an option to thoroughly scientific research techniques in its quest for information and facts.

1.5   Types of Research

The diverse problems and/or question kinds that set research in motion require different research approaches, differentiated by their conceptual or theoretical background and methodologies [4]. In engineering, the different basic *types of research* based on the research study's application, its objectives, and methods of seeking the research information and data include the applied, action, fundamental, experimental, descriptive, and evaluative research, and the case study.

1.5.1   Applied Research

Applied research is a research type, generally carried out to examine the theoretical concepts or solve immediate challenges in real-world problems to improve a process, system or product [7]. Academic research falls into this category because it tries to generalise teaching and learning situations and acts as instructional materials. Applied research in engineering are research, which apply examined theoretical concepts (or scientific principles) for the benefit of humanity. The predominant objective of applied research is to establish a solution(s) for each compelling challenge in real-world practice.

1.5.2   Action Research

Action research is a type of research designed towards the immediate application of generalisations, principles, or theories and solutions to current challenges in a local or global setting. The research outcomes in action research, for instance, are of regional or international application, which a researcher or a team of researchers can direct to alleviate conditions and practices.

1.5.3   Fundamental Research

Fundamental research, also called *pure* or *basic research*, is a type of research concerned with the development and refinement of generalisations, principles, or theories and the development of a novel idea or theory [7]. Research that deal with natural phenomena or are associated with pure mathematics is an example of fundamental research. The objective of pure research is to seek information for advancing knowledge that might have a wide range of applications in the medium-to-long term [7]. It is most closely related to the laboratory conditions and controls typically employed in scientific and engineering research.

1.5.4 Experimental Research

In simple terms, *experimental research* investigates what would likely occur when specific parameters are controlled or manipulated. It is conducted to determine the truth about a process, system, or product under verified conditions, or evaluate the effects of different parameters. It entails isolating a parameter while controlling or manipulating the factors that can impact it.

1.5.5 Descriptive Research

Descriptive research, also known as *expository research*, comprises comparative and correlational techniques as well as fact-finding investigations, to effectively record, explain or describe, analyse and interpret parameters or conditions of an existing system or circumstance [7]. Also, it includes attempts to determine causes, though parameters cannot be manipulated or controlled, but a researcher or a team of researchers can report it. Descriptive research entails comparison and contrasting, and establishes a relationship between parameters, which is conducted to showcase the common or parametric conditions, relationships, and physics of behaviours.

1.5.6 Evaluative Research

Evaluative research is a research designed to compute parameters with relevance, the usefulness of yet to be taken actions or actions that have been implemented.

1.5.7 Case Study

A *case study* is a type of research that deals with the thorough investigation of specific parameters, conditions of a system, or circumstances in an effort to evaluate the effects of a certain factor on it and what assists in its success or failure for in-depth analysis [1].

1.6 Research Methodology

The research methodology is the science of systematically resolving the research problem and/or answering the research questions [8]. Often recognised as how a research is to be methodically conducted, research methodology includes the overall decision on the *research approach* that would be adopted to study a topic, informed by the philosophical assumptions a researcher or a team of researchers bring to bear in the study, inquiry procedures (termed *research design*) and particular *research methods* or *techniques* for collection of data, their analysis and interpretation [9]. Research methodology aims at employing the right procedures for a resolution to the research problem and to create the enabling circumstances for the proper execution of the research methods for collection of data, their analysis and interpretation. It is the research manual, and in itself is a science. For instance, in a topic entitled, '*managing risk factors of injuries*,' the research methodology would involve accident investigation and preventive measures, collation of associated data within the subject matter and approaches that would be involved in the critical data analysis and the like.

Often, these two terms—*research methodology* and *methods* are confused as the same, and strictly speaking, they are not, due to their numerous distinctions. One of the main distinctions between research methodology and methods is that research methodology refers to more than not only a set of methods but also the rationale and philosophical assumptions underlying the particular study. In contrast, research methods are the systematic codified series of steps taken to conduct a specific task on a research topic or to attain specific objectives. In connection to research, research methodology is entrenched in the method's existence while research method is the crucial line of carrying out the research.

In terms of application, research methodology elucidates what and how suitable specific means and techniques would be to a particular topic. In contrast, research methods comprise the approaches that allow studies to be successfully initiated, carried out and established. About the stages involved in research, research methodology is utilised at the start of the research investigations or experimentations to justify the chosen methods, their purpose and how they would serve their functions. At the same time, research methods are more valuable in the latter part of an investigation or an experiment due to their utilisation in drawing proper inferences. In conclusion, the research methodology can be described as a broad multidimensional term, while research methods constitute only a part of the research methodology.

1.7 Research Approaches

The research approach is a plan and the procedure for research that includes stages from far-reaching philosophical assumptions to specific techniques for the collection of data, their analysis and interpretation [9]. The plan comprises numerous research decisions, including the proposed research methodology and discussion of findings. The primary considerations in selecting an appropriate research approach often depend on the nature of the *research problem, questions* or *concerns* being contemplated, the researcher or team of researchers' personal experiences, and the research tools and techniques for the study. The different basic approaches to research are *qualitative, quantitative* and *mixed methods* [9]. Typically, the distinctions between the qualitative and quantitative research approaches are defined using qualitative phenomena, measured or expressed by the quality of something (e.g. examining the reasons for human conduct—why people think or do specific things) rather than quantitative phenomena, measured or expressed in terms of quantity; qualitative words rather than quantitative numerals; and qualitative open-ended problem/questions (e.g. expert focus group interview) rather than quantitative closed-ended problem/questions (e.g. hypotheses testing).

A broader fashion to examine the degrees of distinctions between them is embodied in the philosophical assumptions, research strategy types (e.g. qualitative case study and quantitative survey or experiment), distinct techniques used in carrying out these strategies (e.g. qualitative data collection through observations and quantitative data collection through instrumentations and questionnaires) and unit of analysis described in terms of quality such as text and image data for qualitative approach and quantity such as frequencies, degrees and intensity for quantitative approach. Besides, both approaches have a historical development. In engineering, the quantitative approaches dominate the research forms since the nineteenth century, and the interest in qualitative approach and, in particular, the development of mixed methods, is, at present, on the rise.

1.7.1 Qualitative Research Approaches

A qualitative research approach is an approach to research inquiry, concerned with the subjective assessment of feelings, opinions, or behaviour to explore and understand the meaning an individual or group attributes to techno-socio-environmental or human challenges, expressed by the quality of something [9]. It is a process of research, which consists of evolving problems and procedures, data collection—usually gathered in the setting of the participant, analysis of data—inductively built from details of common topics, and data interpretation—interpreting, commenting, and explaining the implications of the data by the researcher or a team of researchers. The final report structure of a qualitative research approach is flexible, and its inquiry form supports a manner of conducting research, which honours an inductive style and focuses on personal meaning and the significance of solving the situational complication.

1.7.2 Quantitative Research Approaches

A quantitative research approach is an approach to research inquiry, concerned with objective testing of generalisations, principles or theories by exploring the interdependence of parameters, measured commonly using instruments such that numbered data can be computed or collected and analysed [9]. These numbered data could then be analysed using descriptive and inferential statistics, clustering, etc. The set structure of the final report for a quantitative research approach comprises the introduction, theories and literature review, methodology, results and discussion, and conclusions and recommendations. Unlike the qualitative research approach, this form of inquiry is based on the philosophical assumptions that allow it to deductively test theories, build safeguards against bias, control for alternative descriptions, and be capable of generalising and replicating the outcomes.

1.7.3   Mixed Methods Research Approaches

A *mixed methods research approach* is an approach to research inquiry, concerned with gathering and integrating qualitative and quantitative data using different designs, which might include philosophical assumptions and conceptual or theoretical frameworks [9]. The predominant presumption of the mixed methods approach is that it delivers a more comprehensive understanding of the research problem and/or questions than any of the approaches when used alone.

1.8   Research Designs

Research design is an inquiry type within qualitative, quantitative, or mixed methods approaches, providing a particular procedural direction or strategies of investigation [9]. There are no blueprints for planning research, but the purposes of the research determine the design and methodology to be adopted for the research, governed by the idea of '*fitness for purpose*' [10]. The basic research design types are *qualitative*, *quantitative*, and *mixed methods* research designs. However, in planning research, a researcher or a team of researchers chooses not just a *qualitative, quantitative,* or *mixed methods* research to carry out the study. They decide on the categories of research among the research design types.

1.8.1   Qualitative Research Design

Case studies are one of the possible means of conducting *qualitative research design* in engineering. A *case study* is an inquiry design in which a researcher or a team of researchers develops a comprehensive case analysis, which in many instances is a process, program or event analysis. These cases are bounded by activity and time, and the collection of data is conducted by applying a variety of procedures within the sustained time duration.

1.8.2 Quantitative Research Design

A *quantitative research design* is categorised into *experimental research designs*, such as simulation, experimental, structural equation modelling, hierarchical linear modelling, and logistic regression research; and *non-experimental research designs*, such as surveys. The focus of engineering research is on three of these designs, which include simulation, experimental, and survey research.

1.8.2.1 Simulation Research Designs

S*imulation research* involves the artificial design of an environment within which a researcher or a team of researchers can generate relevant data through experimentation trials that allow the examination of a process or system at a steady state or dynamic behaviour under manipulated or controlled scenarios. In engineering applications, the term simulation is referred to as the development of a physical, numerical, or computational model, which depicts the configuration of a system in a steady state or dynamic process. Simulation approaches are beneficial in the development of models for studying and predicting present and future occurrences. With known initial conditions, parameters, and exogenous variables, these simulation models can be implemented to embody the behavioural process of a system.

1.8.2.2 Experimental Research Designs

The e*xperimental research* attempts to establish if a particular treatment impacts an outcome. A researcher or a team of researchers conducts experimental research by executing different treatments on two or more setups or groups, and afterwards evaluates the performances of the setups or groups regarding their individual outcomes. An experiment is distinctive by the much greater control researchers have on the study environment, and in some cases, several parameters might be manipulated or controlled to observe their influence on the other parameters.

The *classes of experimental research designs* include the pre-experimental, true experimental, quasi-experimental, and correlation and ex post facto designs. The *pre-experimental design* is the primitive, undependable experimental method wherein philosophical assumptions are made notwithstanding the non-existence of key parameters' control [4]. The *true experimental design* is an experimental method involving the subjects' random assignments to treatment conditions. The *quasi-experimental design* is an experimental method involving the subjects' nonrandomised assignment. The *correlation design* is an experimental method comprising the search for cause-and-effect relationship between two data sets, and the *ex post facto* design is an experimental method that changes experimentation in the reverse direction, attempting to interpret the observed effects on the basis of the nature of the cause(s) of a phenomenon [4]. The correlation and ex post facto research designs culminate in conclusions that are challenging to substantiate or prove, and so, they rely to a large extent on logic and inference.

1.8.2.3 Survey Research Designs

Survey research seeks to determine the numeric or quantitative description of trends, attitudes, or opinions of a studied (questioned or observed) population sample [8, 9]. It comprises longitudinal and cross-sectional studies employing structured or expert interviews or questionnaires for the collection of data with the intent of postulating generalisations from a population sample. The purpose of survey research is to form a database, e.g., a studied population sample, i.e., observed or questioned, from which characteristics or relationships of the population are then inferred.

1.8.3 Mixed Methods Research Design

The *mixed methods research design* is an inquiry design, concerned with integrating the qualitative and quantitative research designs and data in a research study [9]. There are many designs of mixed methods. The key models are the convergent parallel, explanatory sequential and exploratory sequential mixed methods. The *convergent parallel mixed methods design* is a mixed methods design form wherein a researcher or a team of researchers merges or converges qualitative and quantitative data to conduct a broad analysis of the research problem and/or questions. Typically, the researcher gathers data from both design forms almost side by side and afterwards, integrates them for overall data analysis and interpretation. In this design, inconsistencies or contrasting research outcomes are explained or further investigated.

Explanatory sequential mixed methods design is a mixed methods design form wherein a researcher conducts quantitative research, data analysis, discussion of the findings and afterwards builds these findings to elucidate them in complete detail with qualitative research. Explanatory sequential mixed methods design is termed explanatory due to the further explanation of the initial quantitative results with qualitative data, and sequential because the qualitative stage follows the initial quantitative stage. *Exploratory sequential mixed methods design* is a mixed methods design form in which a researcher conducts qualitative research, data analysis and afterwards builds quantitative research through the qualitatively obtained information. The qualitative stage can be utilised to create instruments, which best fit the study sample under consideration, and identify the suitable instruments or stipulate the required variables to be utilised for the follow-up quantitative study.

Table 1.1 Range of possibilities of qualitative, quantitative and mixed methods

Qualitative methods	Quantitative methods	Mixed methods
Emerging methods	Predetermined methods	Both predetermined and emerging methods
Open-ended questions	Closed-ended questions or instrument-based investigations	Both open- and closed-ended questions, and/or instrument-based investigations
Focus group interview data, observation data, document data, etc.	Performance data, observational data, measurement data, etc.	Multiple forms of data drawing on all possibilities
Text and image analysis	Statistical analysis	Statistical, text and image analysis
Themes, patterns interpretation, etc.	Statistical interpretation	Across databases interpretation

Source Creswell [9]

1.9 Research Methods

Research methods, also called research techniques, refer to a range of approaches employed in research for gathering data that would be utilised as a basis for inferences and interpretation, and for explanation and prediction [10]. Research methods can be expressed as those methods or techniques employed by a researcher or a team of researchers for conducting an inquiry or research, which is aimed at finding solutions to the research problem and/or answers to the research questions. Although research techniques in engineering are limited to the more generic methods that a researcher or a team researchers employs, they might comprise more exact features of the engineering science enterprise, like formulating concepts or theories, developing models and devices or equipment, and conducting experimental and survey studies.

The basic types of research methods are *qualitative, quantitative,* and *mixed methods* research methods. For research methods to be beneficial, a full range of possibilities for collection of data and their organisation should be considered, i.e. qualitative, quantitative, and mixed methods, through their degree of predetermined nature, usage of open-ended against closed-ended problem solving or questioning, and focus on nonnumeric against numeric analysis of data. Table 1.1 presents the range of possibilities for qualitative, quantitative, and mixed methods for engineering researchers.

1.10 Phases in Conducting Research

The *phases in conducting research* include the identification of research topic or problem, conduction of literature search and review, specification of the research purpose, statement of the research problem and or questions, stipulation of the conceptual framework—typically a set of hypotheses (where necessary), choice of methods (for collection of data), data collection and their analysis and interpretation, research reporting and evaluation, and communicating the research findings. Generally, these phases culminate in the overall *research process* involved in conducting research. Still, the researchers should view them as a developing iterative process as opposed to a fixed set of stepladders. Nearly all research starts with a statement of the research problem or perhaps, the objectives of the study, followed by a literature review—a review of theories and literature to identify the knowledge gaps in previous research, providing justification or motivation for the research. Typically, a review of related works in the literature is carried out in the subject area ahead of a *statement of the problem*, and or *research questions* formulation, then, the formulation of the hypothesis—the assumption to be tested; and data are collected, analysed and interpreted using a variety of statistical methods.

1.11   Research Processes

A *research process* comprises a series of actions or stages and their desired sequences, required to conduct research successfully. The typical steps involved in research, in particular, engineering research as a process are the *identification and formulation of the research problem and or questions* in a subject area of research interest, which involves the breaking down of a specific problem into sub-problems, and or research questions are raised on this problem. Also, typical steps involved in the research include research hypotheses, formulated to guide the study (if applicable); *conduction of literature search and review* on the research problem or questions, used to build the research theoretical or conceptual framework; and the *methods for data collection*. Other typical steps involved in the research are the *analysis of collected data and their interpretation*, and *drawing of inferences or conclusions*, which would answer the research questions or lead to solutions to the research problem, and the *writing of recommendations or suggestions*, conducted for further actions, redefining or modifying the problem.

1.12   Chapter Summary

The predominant objective of any engineering research is to establish innovative knowledge based on the principles of scientific thought and measurement for the benefit of humanity. Though each study has its specific purpose, engineering research is performed for the purposes of exploration, description, clarification, understanding, prediction, diagnosis, and experimentation.

The engineering research process starts with an in-depth description of the specific area of research interest, including problem, questions, or phenomenon, which a researcher or a team of researchers wishes to address, seeking solutions, answers, postulation of theories, generalisations, or principles, and applying the scientific solutions, postulation of theories, generalisations, or principles.

In selecting an appropriate research approach, the primary considerations depend on the nature of the research problem, questions or concerns being contemplated, a researcher or a team of researchers' personal experiences, and the instruments for carrying out the study.

References

1. Lues, L., & Lategan, L. O. K. (2006). *RE: Search ABC* (1st ed.). Sun Press.
2. Thiel, D. V. (2014). *Research methods for engineers.* Cambridge University Press.
3. Oyebanji, J. O. (2004). Research and research philosophies In H. A. Saliu & J. O. Oyebanji (Eds.), *A guide on research proposal and report writing* (Ch. 1, pp. 1–8). Faculty of Business and Social Sciences, Unilorin.
4. Walliman, N. (2011). *Your research project: Designing and planning your work.* Sage Publications Ltd.
5. Research. (2018). Forms of research, In *Research.* Wikipedia the Free Encyclopedia. Retrieved from https://en.wikipedia.org/wiki/Research
6. Oyesoji, A. A. (2018). Research conceptualization In C. T. S. Sibinga (Ed.), *Ensuring research integrity and the ethical management of data* (Ch. 10, pp. 174–192). IGI Global.
7. Deb, D., Day, R., & Balas, V. E. (2019). *Research methodology: A practical insight for researchers.* Springer Nature Singapore Pte Ltd.
8. Kothari, C. R. (2004). *Research methodology: Methods and techniques.* New Age International (P) Ltd.
9. Creswell, J. W. (2014). *Research design: Qualitative, quantitative, and mixed methods approaches* (4th ed.). SAGE Publications Inc.
10. Cohen, L., Manion, L., & Morrison, K. (2007). *Research methods in education* (6th ed.). Routledge—Taylor and Francis Group.

Chapter 2
Research Problem and Questions

Abstract The objectives of this chapter are to

(i)   Describe the research problem and questions;
(ii) Identify appropriate research problems and questions;
(iii) Specify the different sources for research problems;
(iv) Enumerate the criteria for selecting a problem for research; and
(v)   Describe the statement of problem.

Keywords Researchable problem · Research question · Research problem selection · Problem statement · Phrasing research questions

2.1 Introduction

Stating the *research problem* and or phrasing *research questions* is a technique for concentrating on a proposed research project. The focus of all research efforts, geared towards proffering solutions to the research problem and answers to the research questions, is at the root of all research projects. Hence, there would always be a research problem or questions, stated to give grounds, reasons, or motivation for the research, and the investigations are designed and conducted to proffer solutions to the research problem and answers to the research questions. This is the target that has to be accomplished before a researcher or a team of researchers can deem any problem or question fit for consideration. Thus, the question is what constitutes a researchable problem.

A researchable problem might arise from the interaction of two or more considerations, resulting in one or more of many promising outcomes, at a confounding state; an undesirable consequence or conflict, which its insightful and actionable strategy is disputable; and so on. Following the statement of the research problem and or research questions phrasing, finding solutions to the research problem and answers to the research questions, classifying the confounding state, eradicating or improving undesirable consequences, or resolving conflicts could afterwards be undertaken.

H. A. Ajimotokan, *Research Techniques*, SpringerBriefs in Applied Sciences and Technology, https://doi.org/10.1007/978-3-031-13109-7_2

2.2 Research Problems

The research problem is the questions or challenges that the proposed research is posed to answer or solve to fill the knowledge gap in existing studies or contribute to the existing knowledge body in the study area. Generally, a research problem can be referred to as a specific issue, difficulty, or challenge that a researcher or a team of researchers experiences and wants to solve in the context of either a theoretical or practical situation [1]. It provides the focus of the research project, culminating in the background study, and the initiator of the specific research tasks, which should be succinctly described to elucidate the nature of the problem and the rationale for its significance [2].

The research problem may be stated in abstract terms, either in a *declarative, descriptive,* or *question form*, however, through the statement of sub-problems, researchers must express how the problem could be rationally investigated [2]. Stating the research problem in question form may assist the researcher in breaking the main problem into sub-problems for better focus and understanding.

2.3 Identification of Research Problems

Researchers, in particular, typically first-timers or beginners, might find it very challenging to choose an appropriate research problem, and they might expend substantial time on examining the numerous research problems without being capable of deciding on which is suitable to select. Common errors in identifying research problems occur when circumstances, objectives, hypotheses, or difficult thoughts are misunderstood as problems. A research problem is distinct from the problem area of interest, which is a broad topic, event, phenomenon, or subject area to be studied. There are three *systematic stages of reductive inference*, essential to explore the problem area of interest to arrive at a research problem or topic. These involve *selecting a problem area* in the field of interest, *pruning the problem area* of interest down to a size that is manageable, and *expressing it succinctly* in an empirically researchable manner. These stages might not be as easy-to-use as specified because researchers do not find it easy to state a fit-for-purpose research problem without thorough and logical deductive inferences.

However, if personal interests guide one to identify a research problem or topic that appeals to a researcher or a team of researchers due to their prior readings, undertakings or interactions, the researchers would be enthused to expend the lengthy hours and painstaking efforts, required to carry out the study, effectively and successfully. Moreover, if there are strong convictions about the need to proffer solutions to a problem, it becomes easier for a researcher or a research team to successfully conduct a good research. Note that such good research must be capable of contributing a unique and meaningful contribution to knowledge.

2.4 Sources of Research Problems

The different sources for the *statement or formulation of the research problem* include personal experiences, literature, experts' opinions, Internet sources, replication, theories, and government publications [3], among others. It is unethical and fraudulent to plagiarise an existing project topic and its problem statement from a written project report because it impedes and or hinders academic growth and research and technology developments.

2.4.1 Personal Experiences

Experienced tutors or those in academia and industry with a long experience or observed existing practises in the academia or industrial sector would know or be familiar with numerous imperfections, inconsistencies, puzzles, or several knowledge gaps requiring solutions. Moreover, through professional experiences or interactions with people and facilities, researchers can find areas where the knowledge gap exists.

2.4.2 Literature

When an in-depth literature review is conducted, it offers several researchable problems. Dissertations, theses, textbooks, and the communicated research findings in journals and conference proceedings, among others, present numerous information on researchable problems. Besides, a researcher or a team of researchers may encounter recommendations for further research work, which can be obtained in dissertations and theses, some inconsistencies, and disagreeable findings in a number of investigated areas. Consequently, research can be carried out to fill the knowledge gaps. Moreover, any study can contribute to knowledge by enhancing the methodology or modifying the existing generalisations, principles, theories, or research.

2.4.3 Experts' Opinions

Consultations with engineering experts, academics, research fellows and thesis supervisors, among others, can aid in the identifitication a research problem. This would help to clarify the researcher's thinking to attain a sense of focus and be articulate and concise in the statement of the research problem, which depends on the researcher or team of researchers' interest in inventing an innovative thought problem.

2.4.4   Internet Sources

Nowadays, e-learning—a piloted learning, usually over the Internet through electronic media—is one of the most reliable sources of getting current and up-to-date
information in any engineering discipline. Through the Internet, a researcher or a
team of researchers can take advantage of the opportunities or available resources
to get up-to-date research outcomes or reports on any area of interest. Thus, the
Internet presents access to different methods that can be employed for solving the
same problem, making it a good source of research problems.

2.4.5   Replication

The *replication* or *repetition* of previous research in the literature could assist in
increasing the findings' validity and generalisability, whose relevance or technical
requirements need research, making them good sources of a research problem.
However, researchers may conduct this replication by exploiting different methodologies, instruments, geographical settings, time frames, or case studies.

2.4.6   Theories

A *theory* is a statement that suggests an explanation of phenomena, events, or circumstances. Theories postulate general principles that their relevance or technical circumstances need research, making them good sources of a research problem because relationships between parameters are expected to be predicted, tested, and established
through theories.

2.4.7   Government Publications

Sometimes, government policies, intentions, or opinions are made public via government publications such as memos, circulars, whitepapers, gazettes, newspapers, or
radio and television ads. Research problems can originate from proffering solutions
to these governmental identified challenges.

2.5   Criteria for Research Problem Selection

Often, there are numerous promising researchable problems due to the fact that there is no shortage of them when well-thought out the world, but the challenges are in selecting the most practical problem for scrutiny among the identified research problems. Researchers, in particular, first-timers or beginners, experience this problematic situation. So, they make attempts or trials on several research problems and leave them for an alternative before selecting one, causing needless delays in conducting their research. To prevent these drawbacks, researchers should examine the available problems to choose the utmost expedient for research. However, for a problem to be researchable, it needs to possess some essential features [2]. These features include significance, researchability, delineation, suitability, and viability [4].

2.5.1   Significance

Deepening knowledge is the fundamental goal of research, and it is appropriate to select a research problem for which the solution would make one of the utmost significant contributions to knowledge. The area of these contributions might be in the methodology, theories, or findings replication to generate a better understanding or enhance the findings' validity and generalisability.

2.5.2   Researchability

For any difficulty, issue, or challenge to be a researchable problem, it must comprise definable and measurable parameters. Researchers must note that they cannot subject all difficulties, issues, or challenges to a systematised investigation or empirical study, such as many metaphysical and ethical challenges.

2.5.3   Delineation

Considering the available time to complete a research and the depth to which the research problem would be addressed, the problem should be capable of being delineated within the subject matter, in particular, to cover a restricted field for a more thorough study or a wide field for a less thorough study.

2.5.4 Suitability

Research problems to be investigated must meet specific research peculiarities. For an issue, difficulty, or challenge to be a problem that is suitable for research, it must meet these specific research peculiarities as follows:

(i)   It should be relevant to a researcher or a team of researchers' technical and professional goals, advancing their knowledge and proficiency in their career.

(ii)   It should be meaningful and exciting, enthusing the researcher or a team of researchers to persevere and investigate the problem exhaustively from the beginning of the research to the end.

(iii) The solution is expected to be within a researcher or a team of researchers' level of competence, knowledge of using relevant tools and techniques; else, they must learn the knowledge within a realistic time frame. That is, the researchers must possess the appropriate skill set and competencies, and have good knowledge of the conceptual frameworks and existing theories in the problem area.

(iv)   The needed equipment, fund, and other resources should be available or accessible. Researchers should shun research problems with numerous parameters that a simple, large-scale study by a research team, having access to adequate funding, can tackle.

(v)   The required time to attain a suitable solution to the problem should be real. Researchers must consider this time limit in selecting a relevant problem because there is always a time frame for research, in particular, those conducted for degree purposes.

2.5.5 Viability

A research problem should be viable, expandable, or able to be followed up with further research. As a researcher or a team of researchers proffer solutions to a problem, they could generate difficulties, issues, or challenges that require further investigations.

2.6 Statement of Problem

The *statement of the research problem* is the definition of what the research to be conducted proposes to do, showing the existing gap in knowledge that a researcher or a team of researchers intends to fill. The statement of the problem must be succinct, precise, and an informative and convincing assertion of the subject matter and the planned parameters for investigation. It must be definite, a matter of fact, and should be presented in a logical sequence, starting with information or theories necessary for problem comprehension, some justifications that may include citations and a declarative or descriptive statement, or intensification in question form. The problem statement is expected to be structured such that it comprises the predominant *significance of the research area,* the *established knowledge gap,* and *the assumed solutions.*

2.7   Research Questions

Typically, it is always a challenging task for any researcher or a team of researchers to design a single, concise question or to phrase research questions correctly. Researchers might need to consider some iterations before a research question becomes appropriate. The research questions would candidly result in two or more different investigation methods, which might be categorised into a few research objectives. Research questions may be phrased employing any of these words, namely, *What? Will? How? Why? and When?* For instance,

(i)   What will be the influence on the environment when a solar-biomass hybrid power generator system is used for micro-grid electrification? or
      Will there be any influence on the environment when a solar-biomass hybrid power generator system is used for micro-grid electrification?
(ii) What will be the bridge between electricity demand and supply gaps when a solar-biomass hybrid power generator system is used for micro-grid electrification? or
      Will there be any bridge between electricity demand and supply gaps when a solar-biomass hybrid power generator system is used for micro-grid electrification?
(iii) How prospective can the solar-biomass hybrid power generator system be for micro-grid electrification? or
      Will there be any prospect of using the solar-biomass hybrid power generator system for micro-grid electrification?

Researchers, in particular, first-times and beginners, should note that various methods of approach are advised for an individual research question. Thus, researchers are expected to use two or more methods of investigation in any research with the hope that the results from various approaches might be utilised for substantiating the project conclusions. Though this enhances the robustness of the research outcomes, it also adds confidence to the research findings.

2.8   Chapter Summary

The various sources for the formulation of research problems are through personal experiences, literature, experts' opinions, Internet sources, replication, theory, and government publications, among others. The primary criteria for selecting a problem for research include significance, researchability, suitability, and viability. The problem statement, defined as what the research to be conducted proposes to do; should be structured such that it comprises the significance of the research area, establishes a knowledge gap, and assumes the solution.

References

1. Kothari, C. R. (2004). *Research methodology: Methods and techniques.* New Age International (P) Ltd.
2. Walliman, N. (2011). *Research methods: The basics.* Routledge—Taylor and Francis Group.
3. Pandey, P., & Pandey, M. M. (2015). *Research methodology: Methods and techniques.* Bridge Center.
4. Walliman, N. (2011). *Your research project: Designing and planning your work.* Sage Publications Ltd.

Chapter 3
Use of Libraries, Literature Search and Review

Abstract The objectives of this chapter are to

(i)   Describe the use of the library and specify how to research using it;
(ii) Define the terms literature search and review;
(iii) Outline the importance of literature search and review;
(iv) Specify and briefly describe the sources of archival literature;
(v)   Specify and describe the types of publications; and
(vi) Define the term search strategy and specify the approaches to literature search.

Keywords Archival literature · Literature search strategy · Full paper · Short journal article · Conference paper

3.1 Introduction

The library is the foremost centre of intellectualism, which doubles as the heartbeat of any academic institution and a storehouse of knowledge. Thus, there is a need for library patronisers—engineers, scientists, researchers, etc.—to have an in-depth understanding of how *literature works* and *information* can be accessed and or sourced when using the library. While research conducted using the Internet might be a hefty alternative, it would certainly not supplant the *use of libraries*. The libraries hold books and periodicals, primary and secondary *sources of information* in hundreds of thousands, and trained personnel who can help and guide researchers through the research process. As part of a research process, a *literature search and review* is conducted.

3.2 Using the Library

Using the library is of vital importance to every research and its communication processes because a library is the best habitat for several decades of published *periodicals*, often called *journal articles*. Furthermore, the web-based search alone would always not be sufficient for quality research. The prevalent means to research a topic is by periodicals, through the use of the *guide to periodicals*—a vast periodical index in every library for successive years. It indexes all topics and people ever referred to in a published periodical. Similar to the library books' catalogue, the guide to periodicals could be deployed to find any periodical on an idea or a person. Write the necessary bibliographical information from the subject index, i.e. the article's author, publication year, the title of the article, publisher of the periodical, volume and page number, and then, search for this specific article in the library periodical section.

The periodical sections mostly contain only bound copies of periodicals for recent years, and for those over a decade or more, microfilms are used to read them. Since the periodical volumes are so large, libraries store them on microfilms, where a librarian would set up the machine to read the articles from the microfilm. Afterwards, the researcher might print the pages if the need arises. Researchers must always write down all bibliographical information for each publication used or those they create and store them electronically using a referencing manager.

3.2.1   How to Research Using the Library

Before going to the library,

(i)   A researcher needs to have an idea or know what s/he wants to research. Otherwise, the researcher would waste numerous library hours due to a lack of an idea of what to seek.

(ii) The researcher should go to the library equipped with tools, such as a pen, jotter, and a bag, capable of holding some books, dissertations, and scientific articles, planned for taking home.

At the library,

(i)   The researcher should have a chat with a librarian where possible. The librarians are meticulous tutors who are familiar with every nook and cranny of their specific library and are well trained to assist with research.

(ii) If the researcher does not wish to have a chat with the librarian, then s/he should go straight to a computer (where available), which has supplanted the card catalogues. However, the card catalogue still exists, but on a computer in a more accessible and legible format.

(iii) The researcher should access the online library catalogue or guide to periodicals. In case the researcher gets perplexed, s/he should solicit assistance from a librarian.

(iv) The researcher should search for the proposed topic or subject matter (as you would conduct a web-based search), either by topic, subject, title, author, or any combination thereof. A list of books or scientific articles would appear on the screen.

(v)   The researcher should write down three to five of the catalogue numbers for books by looking at the first two letters of the codes that correspond with a type of discipline-based print materials (engineering, etc.), or the required bibliographical information among those given in the index of the guide to periodicals for that subject so that s/he can search for them.

(vi)   The researcher should search for and then find the book (by catalogue number) in the book section or the scientific papers in the periodical section.

(vii)   Once the researcher has located one or more relevant books or scientific papers, s/he should sit in the library to read them or the required pages. The researcher should take notes on a jotter (where necessary) and ensure that s/he records every piece of bibliographical information (author, publisher, published date, etc.).

(viii) The researcher should make copies of the required pages from the books and their title pages, including all bibliographic information and or the scientific articles, using the copy machine.

(ix)   Check the book(s) out of the library (if necessary).

3.2.2 *Checking Books, Dissertations and Scientific Articles Out of a Library*

To check a book, dissertation, or scientific article out of a library requires a library user or researcher to fill out paperwork to obtain a library card. Generally, library books, dissertations, or scientific papers are loaned to the public for approximately 2–6 weeks, depending on the publication types. The researcher should ensure that s/he obtains from the library the form or card with the due date for the return of the borrowed books, dissertations, or scientific articles. The timely return of a book, dissertation, or scientific paper to the library is significant because late fees can add up, costing a fortune, and also, to guide against a ban from the library due to added up late fees over time. However, make sure that all bibliographical information is written down, or s/he should create and store them electronically using a referencing manager before borrowed books, dissertations, or scientific articles are returned, because citations and listing of references would always be needed during research and communication processes of project report writing, conference and journal paper presentations and publications.

3.3 Literature Search and Review

A *literature search* is an organised, carefully considered, and planned search, designed to identify existing theories, research, and information from the published literature on a subject matter. A well-thought out literature search is one of the most efficient ways to find the available sound body of facts or data (evidence) on a research subject area. This body of facts or data can emanate from various sources, either in print or electronic form. On the basis of the identified research and information from the published literature, often called *archival literature* since it is stored permanently, a review of theory and literature is carried out. A *literature review* is a written piece that summarises and analyses the existing theories, theoretical or conceptual frameworks, research and information found through a literature search. It examines all relevant published information and data on a topic, and takes into consideration their contributions and weaknesses, including the strengths, opportunities, or threats [1].

3.3.1 Importance of Literature Search and Review

Using the published literature is an essential part of any research and its communication processes, because the published literature connects the proposed research studies to broader scholarly knowledge, demonstrates the researchers' understanding, and puts any research done in a broader context. A literature search and review is carried out to [2]

(i)   Provide an academic basis for the proposed research;
(ii) Clarify the researcher or a team of researchers' ideas and research findings;
(iii) Evaluate the various research designs employed as methods of data collection in various studies;
(iv) Discover the methods of data analysis and possible findings; and
(v)   Pin out the possible issues, difficulties, or challenges with the proposed research project.

Several research projects entail doing your studies, while others might involve analysing the primary sources or the literature itself. In any of these cases, the information found during the literature search informs and underpins the research decisions, which include the proposed research methodology and discussion of results.

3.3.2 Sources of Archival Literature

Global knowledge, in particular, in the engineering field, is digitally stored as published books and scientific papers in printed form, typically accessible through the use of a library and web-based search with the appropriate keywords. The sources of this archival literature for information may either be in *print* (e.g. books, refereed journal articles, and conference papers) or *non-print* (non-book)—generally referred to as audio-visual materials (microforms, computer files, films, etc.).

3.4 Literature Search Strategy

A *search strategy* is an organised, well-thought out plan for seeking information from the literature. The literature search strategy is, in particular, essential when utilising electronic citation databases, such as SCOPUS, ASME Digital Collection, or ETDE World Energy Base, as it keeps a researcher or a team of researchers focused on the topic and within the boundaries of what is required to search. However, there are a number of *approaches to conducting a literature search strategy* [3, 4]. These approaches include systematic, retrospective, citation, and targeted searches. The *systematic search* is a literature search that involves the finding and follow-up of all relevant scientific articles, books, and reading lists. The *retrospective search* is a literature search that consists of the finding of the most recent scientific articles and books, and afterwards working back to older reading lists. The *citation search* is a literature search that involves the finding and follow-up of references from valuable scientific articles, books, and reading lists. The *targeted search* is a literature search that consists of the restriction of the research topic to focus on a narrow area of the literature.

In practice, a researcher or a team of researchers uses a mixture of these approaches. This mixed approach might comprise using systematic searching to find all relevant materials; adopting a retrospective approach for finding the most recent material to older ones; using citation search to obtain valuable leads from beneficial articles, books, and reading lists; and embracing a more targeted approach with a focus on a clear picture of what the research requires.

3.5 Types of Publications

The different *types of publications* can be categorised into books, journal articles, conference papers, standards, patents, and dissertations [5], among others. Though journal articles are the most rigorously peer reviewed publication type with credible scientific evidence, a researcher or a team of researchers might discover several innovative ideas in other publication types. Thus, a literature search and review of all publication types should be conducted to underpin the research decisions, and such publications must be in-text cited in the main text and listed in its reference section. The credibility of any scientific evidence can be distinguished from speculations, conjectures, and often objective mistakes, partly through a check of the information publication source. A researcher or a team of researchers must be able to differentiate based on the different publication types, and they should simply employ only archival literature, undoubtedly deep-rooted in the scientific method of peer review. While the peer review processes might not be infallible, they are the best and extensively the most acceptable method, which is utilised to preserve the integrity of scientific principles. Engineering researchers must realise this and know where to publish their research findings.

3.5.1   Books

The book can provide a great start with particular or generic information on a topic or subject matter for a researcher or a team of researchers. The search for up-to-date books—typically, those published within the last one or two decades on a subject area or a particular research topic—can be conducted in an academic or engineering library. The most commonly utilised book types by researchers are *textbooks, research books,* and *reference books* [5]. A *textbook* is a book utilised as a standard work for the study of a specific course or subject. Typically, textbooks have an enormous distribution and sometimes periodic editions as the texts are revised for appropriateness and an up-to-date reflection of the current knowledge, utilised commonly in schools and training institutions for instructions. Generally, they cover theoretical, numerical, and standard experimental techniques employed in the engineering field with detailed references and bibliographies. When writing, researchers do not have to restate an established theory or theories, but instead utilise applicable textbooks as a reference. Textbooks are as well used as a source of definitions, equations, and standard test methods. The *research book* is a book written by a research specialist or specialists in a field for a small targeted audience, such as specialised academic communities, which comprises high-level pieces of information on a subject area or specific topics. The *reference book* is an electronic resource, book or a set of books, providing particular or generic information that is arranged in alphabetical order or an index of terms, typically intended to be consulted for information on specific matters. Examples of reference books are encyclopaedias, dictionaries, etc.

3.5.2   Journal Articles

The refereed *journal article* is of the most vital importance and is a cherished *contribution to the archival literature* and one of the best sources of information because of its predominant ground-breaking research content [5]. Researchers may select journal articles for being specific and recent, and essentially, they should search for research in scholarly peer-reviewed journals because such published articles have undergone peer critiquing—validation and some quality control before publishing. The refereed science and engineering journal articles are best discovered utilising *citation databases*, like SCOPUS, ASME Digital Collection, and ETDE World Energy Base, among others. A *search strategy* might be utilised on all these databases through their unique search features. Numerous databases have '*help screens*' or '*tutorials*' to assist in familiarising with their search interface. A journal articles can either be a *full paper* or *a short journal article.*

3.5.2.1   Full Papers

The *full paper* can be grouped into four primary forms, and it is essential to be able to specify and differentiate between them. These include research articles, review articles, commentaries and opinions, and case studies [5]. The *research article*, including a systematic review, is a scholarly article requiring original research that typically comprises an abstract, background to the study, description of the research details, results and discussion of their relevance, and conclusions. Research articles are the best sources of information because they provide access to the most current, cutting-edge research and sound information on older research. The *review article* is a scholarly article that gives a summary of the research articles on a topic. Though research articles generally report the latest advancements, review articles may not report new knowledge (from an author or a team of authors). Still, they would have an enormous reference list that encompasses up-to-date research and technological developments. They are published to *consolidate knowledge* in the subject area or on a specific topic. The *commentary and opinion* is a scholarly article that expresses personal interpretations, opinions, or innovative perspectives about existing research or a letter to the editor-in-chief on a topic. The *case study* is a scholarly article that focuses on a single situation or scenario, as opposed to a group of various studies.

3.5.2.2   Short Journal Articles

The *short journal article* can take the form of letters, comments, short communications, errata, and technical notes [5]. Some specific journals publish *letters*, requiring an author or a team of authors to submit short articles with page limits, a limited number of words, tables, and figures. Due to these restrictions, letters have few explanations, descriptions, and discussions, and only a few references. Though the process of review is nevertheless rigorous, the reviewers are requested to specify to the editor-in-chief of the journal a decision of a '*yes* or *no*' to publish. If the decision is a yes, the submission would be published without any review feedback from the author or a team of authors. Hence, the publication of letters is much more speedily compared to the full papers. The *comment* on a published full paper is a scholarly article submission by another researcher or a team of researchers who have read the published article and have suggested mistakes, misrepresentation, or misinformation and or an omission or failure of a thorough literature review in the article. Typically, such a comment is conveyed to the original author or a team of authors for a response, and the journal would publish their response in the same journal volume or issue. The *short communication* is limited to the publication of innovative but minor findings and is often utilised as a means to earn the rapid publication of novel research findings. Comments and short communications are both published without a thorough review of the literature. The *errata* are a list of corrected mistakes published in a subsequent journal issue.

The review process of short articles is a quick procedure, involving just the editor-in-chief or associate editor, for some publications, to review the submission. Every scholarly journal has an international standard serial number (ISSN), found on the front page of each journal issue and often the digital object identifier (DOI) of an article on the first page or pages of each journal article, among others. Globally, most names of scholarly journals are unique; hitherto, an author or a team of authors needed to list the correct journal title as numerous journals hold names that are alike.

3.5.3   Conference Papers

A *conference* is a formal meeting of people (with a shared interest) for discussion that is characteristically held over a couple of days. It is a preferred mode to disseminate recent findings for a researcher or a team of researchers because, universally, the sequence of events regarding submission and publication of *conference papers* is significantly more rapid compared to the review and publication process of journal articles and some other publication types. So, the engineering science conference, a formal meeting of researchers in science and engineering disciplines, is for reporting and discussing research and technology developments yet to be published [5]. Conference papers, with the latest findings, are presented and discussed by engineering researchers at the conference through formal or poster presentations, onsite or remotely. Written conference manuscripts, typically requested for submission through a call for papers, are submitted by authors to the conference technical committee. Afterwards, a peer review of the submissions is carried out for relevance and correctness on a pass (with or without editorial changes) or fail basis. If accepted, the author is invited to make an onsite or remote conference paper presentation. The conference paper is published and released to the conference attendees, among others, as *conference proceedings*. Though conference publications are a dependable source of different research and technology developments, they are prone to uncertainty due to a lack of a rigorous review process.

3.5.4   Standards

In engineering, a *standard* is a document that outlines the specifications of a requirement or a particular experimental method [5]. Also, standards are employed to define engineering terminologies, codes, and so on, such that engineering as a profession utilises terminologies in a very appropriate, satisfactory, and well-defined manner. Researchers, in particular first-timers or beginners, are expected to be conversant with these terminologies and their succinct definitions and utilisation when writing scientific articles. These standards are reviewed as research and technology development matures, and as the standard evolves, they are written and experienced professional members of the relevant discipline approve them. Behind every review of an engineering standard, specific modifications are carried out to reconcile uncertainties or confusions and for adding novel terminologies as the technology emerges. Though the International Organization for Standardization (ISO) and the International Electrotechnical Commission (IEC) maintain standards internationally, numerous countries or regions have their own national or regional standard authorities in charge of maintaining and implementing local, regional, and international standards [5].

3.5.5 Patents

A *patent* is an authority or licence granted by the government of a sovereign state, conferring a set of sole rights to an idea to an inventor for commercial advantage and exploitation within a set time, in exchange for an invention's detailed public disclosure to proscribe others from making, using, or selling it [5, 6]. Patents are granted based on their originality, called *inventive step*. The object of patents is to protect ideas, innovations, inventions or research outcomes for commerce advantage against theft such that other person(s), group(s) or company(companies) do not make gains from the commercial business of the invention [5]. Most patents, maintained by each country or the World Intellectual Property Organization through an inventory of the granted patents, can be found by employing a web-based search with the appropriate keywords.

Numerous technologies, in particular, those of science and engineering, are protected by patents, which might constrain their deployment in research and technology developments and businesses. However, neither the patent award implies that the technology works in the defined manner nor that the process or method has merits over current technologies, notwithstanding the patent's documented claims [5]. Hence, when a researcher or a team of researchers conducts a literature search and review, they should be conscious that though a patent contains novel ideas and their innovative usages, still, patents are typically not dependable and verified research and technology development sources. A patent is characterised by its *number, date of submission, authors and their affiliations, sponsoring organisation, summary,* and *a series of six-digit codes*, defining the field in which the invention would find applications.

3.5.6 Dissertations and Theses

Several tertiary institutions require undergraduate and graduate learners to submit a *dissertation*—project report or thesis as one of the final assessments. When a dissertation is successfully defended, it is predominantly stored digitally in printed form, which is often available online. However, dissertation status differs from institutions based on their assessment processes. Mostly, undergraduate project reports are graded on a pass or fail basis without a rigorous content revision according to the given feedback from the examiner. However, they do offer valuable information, but may be an unreliable source of unverified knowledge.

On the other hand, postgraduate theses are customarily revised according to the given feedback of the examiners; thus, they are a reliable information source. Most often, dissertations and theses contain valuable information on numerical, computational, or experimental methods that are abridged in conference or journal papers. Moreover, it is usual for dissertations and theses to comprise circuit diagrams for projects associated with electrical engineering, mechanical drawings for projects related to mechanical and structural engineering, micrographs for projects related to material science and engineering, or details of chemical processes for projects associated with chemical engineering, among others.

3.5.7 Internet

The *Internet* source is an extremely valuable information source with innumerable research and technology developments and their published statistics, in particular, those sponsored by academics, academia, industries, governments, and non-governmental organisations. Valuable information, like government policies, infomercials, and standards, is accessible and often available for download in full at no cost. The Internet is one of the excellent sources for finding official publications in full. Sites like the US Department of Energy (http://energy.gov) provide links to government documents. Statistics are fundamental elements of a research and are generally more accessible using the Internet. However, be very cautious in evaluating sites because any piece of information can be put on the Internet by anyone.

3.5.8 Infomercials

An *infomercial* refers to any printed article or video clip made by businesses to sell their products and expertise [5]. These printed articles contain several facts, specifications, and information but with a little or near absence of the employed research and technology development approach for product development. The claims of the articles are unsubstantiated because they are not subjected to independent experts' rigorous review. Infomercials are found in trade journals, on websites, etc., which may bear a resemblance to scientific or newspaper articles. Still, researchers might typically label them succinctly as not part of the standard technical content.

3.6 Chapter Summary

The use of the library and scholarly web-based search may offer quick access to the archival literature, which is always ordered according to its date of publication and disciplines. A literature search and review is an essential component in the design and implementation of a research project. The refereed published scientific papers, characterised by a distinctive title, author or authors' list and their contact details, submission and acceptance dates, an abstract that provides a complete synopsis of the research project and keywords, among others, are the best sources of dependable engineering science information. In conducting a literature review, a researcher or a team of researchers is expected to read published books and scientific papers to identify relevant existing theories, research, and information, which, in particular the methodology and research inferences, are reported in a brief and clearly expressed manner. The succinctly written summary should include the book or scientific paper's full reference, employed methodology, and the significant research findings and conclusions from the research project.

Approaches to literature search include systematic search—finding and follow-up of every relevant scientific article, book, and reading list; retrospective search—finding the most current scientific articles and books, and then working back to older reading lists; citation search—finding and follow-up of references from useful archival literature; and targeted search—restricting the research topic to focus on a narrow area of the literature.

The different types of publications include books, journal articles, conference papers, patents, and dissertations, among others. Journal articles are the most rigorously peer reviewed publication type with credible scientific evidence. However, a researcher or a team of researchers is expected to conduct a literature search and review of all publication types because they might discover several innovative ideas in the other types.

References

1. Lues, L., & Lategan, L. O. K. (2006). *RE: Search ABC* (1st ed.). Sun Press.
2. Walliman, N. (2011). *Your research project: Designing and planning your work.* Sage Publications Ltd.
3. Bell, J. (2010). *Doing your research project: A guide for first-time researchers in education and social science* (5th ed.). Open University Press.
4. Baker, S. (1999). Finding and searching information sources. In *Doing your research project: A guide for first-time researchers in education and social science* (Ch. 5, 3rd ed.). Open University Press.
5. Thiel, D. V. (2014). *Research methods for engineers.* Cambridge University Press.
6. Patent. (2019). Patent, Wikipedia the Free Encyclopedia. Retrieved from https://en.wikipedia.org/wiki/Patent

Chapter 4
Developing a Research Plan

Abstract The objectives of this chapter are to

(i)   Describe the terms research proposal and research protocol;
(ii) Specify and discuss the elements of research proposal;
(iii) Specify the goals of research protocol;
(iv) Outline preferable sequence for the different section headings of a research protocol and discuss their contents; and
(v)   Discuss the basic engineering research tools and techniques.

Keywords Research proposal · Research project timeline · Research project budget · Research protocol · Research tools and techniques

4.1   Introduction

Primarily, the focus of developing a research plan is to assist researchers with their engineering research project design or planning and guide them in the compilation of the required *research tools* and *techniques*. Often, these tools and techniques differ from one engineering discipline to the other. The basic approach to designing a research project is through a *research proposal*, which should comprise five sub-divisions—*project title, executive summary, the main body* of the proposal, *project timeline, and* the *budget*. The *main body* of the proposal includes the *introduction, literature review, methodology,* and the *preliminary results* (if any), with well-established subsections in each section. Therefore, most research studies begin with a written research proposal, generally demanded by universities and research institutions from their prospective research students or fellows, in particular, from prospective MPhil or PhD students, academic fellows, or those competing for funded research projects, locally or internationally.

For a researcher or a team of researchers to be effective and concise, a research project requires a road map, termed the *research protocol*, whose details should ensure that the researcher would successfully execute the research. Developing an appropriate research protocol at the first attempt might prove very challenging. However, it is sensible first to get better clarity on the research topic and the research project design ahead of the research protocol drafting. The researchers should seek clarification on the topic and what the focus of the research would be. A simple approach to formulating a research topic can be through a researcher's consultations with prospective supervisor(s) and experts in the field, and then use the libraries, Internet facilities, consulting databases like the NEXUS for current and completed research projects and other resources like project reports, conference proceedings, journals, and books, among other valuable resources.

4.2 Writing a Research Proposal

A *research proposal* is a written document, concerned with a comprehensive description of a proposed research plan or programme on a specific subject matter or topic to substantiate the need and relevance of carrying out the research [1]. Research proposals should draw attention to the proposed study's benefits and possible *research outcomes*, backed by informative and convincing evidence. They are merely a framework of the whole research process, which provides the readership of the proposal with a synopsis of the research project to be conducted, information, and persuasive evidence to be discussed. Because a research proposal is designed to be read by different people—not only expert fellow researchers, but also non-experts in the subject area—the language employed for the description of techniques should be as non-technical as possible.

All research proposals have roughly the same format, be it an employee opinion, health care, review work, or research and technology development projects, and are written in present and future tenses. They all have the same basic structure and format; however, they can take several forms. Some might be detailed—running to ten pages or more—while others might be precise and straight to the point (with less than ten pages). Considerably, this depends on the details of the description of the different components. The rationale is that the readership of the research proposals (i.e. academics, financiers, funding agencies, and so on) are generally aware of the sections where to search for the essential information required, irrespective of the form the research proposal takes. Whatever the form it takes, numerous steps are required in writing a good and robust research proposal, with each action being treated systematically and adequately. Research proposal writing is carried out for numerous purposes, such as to seek a research grant, as a requirement to obtain ethical certification for a research, as an initial task ahead of undertaking a research project in tertiary education, and as a condition for a job offer in research-based institutions (that typically need sponsor-approved research proposals), among others.

Numerous primary concerns are necessary before designing or planning a research proposal. These concerns include selecting an appropriate research approach after better clarity on the topic of the research, reviewing theories and literature to situate the planned research within the existing literature, and engaging from the inception of sound writing and ethical practices. The style of a piece of proposal writing has lots of similitudes to those of scientific papers. However, a research proposal is written in future tense with distinctive points of emphasis. Similar to scientific papers, a research proposal has different sections, including the background to the study, problem statement, aim and objectives, justification, literature review, methodology, project timeline, budget and references.

Typically, a research proposal should consist of a broad but focused *literature review*; a *research methodology*, far more comprehensive than that of a scientific paper, permitting a deeper understanding of the research worth, its hazards, and strategies for mitigating them; and a section describing the *expected research outcomes* or *hypotheses*. Also, it might include *preliminary results* (if available) and, typically, the curriculum vitae of the researcher or a team of researchers. The curriculum vitae are necessary to attest that the researchers seeking to carry out the research are capable. Research funders announce calls for proposals, stipulating the research topics to be funded and often requiring a detailed format. These funders might be government funding agencies, non-profit organisations or business research foundations.

4.2.1 Elements of a Research Proposal

The following suggested outline of a research proposal is deemed fit as the structured approach to writing dissertations, theses, or reports. Nonetheless, the format can be adjusted and personalised to suit the needs of organisations soliciting one. The essential *elements* or *sections* of a research proposal include the *project title, executive summary, project description, project timeline, budget,* and *references*.

4.2.1.1 Project Title

The research *project title* must be unique, specific, and succinctly descriptive to communicate the research direction and discipline or field. It occupies the first page of a research proposal—a page, which should contain the title of the project, the author's name, the proposal's relationship to a research project or degree requirement, the institution name where the proposal is to be submitted, and the presentation month. The project title must be expressed in such a manner that encapsulates, in a few words, the essence of the proposed research and is not stated in such a broad style, which claims more than it essentially can deliver. The use of acronyms in the project title is not recommended because the reader or assessing panellist may not be familiar with them.

4.2.1.2 Executive Summary

An *executive summary*, also known as a summary of the proposal or project summary, is a synopsis of the research proposal that entails an overview of the *background to the study*, the *main objective* of the research, a brief description of the *approach to the study,* and often, the *research implication,* i.e., a brief and clearly expressed statement on how the project would be beneficial and fit into the research field goals and objectives. Also, *keywords* should be included under the executive summary for an easy and thorough search of the literature (in the field of competing for research and technology development outcomes).

4.2.1.3   Project Description

The research *project description,* also called project details, comprises the *introductory, literature review, research methodology,* and *expected research outcomes* sections. Like in scientific articles, the *introduction* or *background of the study* section, is the first step in the research proposal writing, which entails a summarised review of the important background information, giving a situational assessment of the subject area or the project theme in the area of specific research interest. This succinct situation assessment should involve a systemic analysis of the past and current information, and the exciting, present and future changes, to identify trends, forces, and or conditions with the possibility of influencing the purpose of the study and the selection of suitable approaches to the study. It should also cover the socio-economic, environmental, geographical, and political concerns within the subject area or in the area of specific research interest.

There are four primary goals of the introductory section in a research proposal. These are to

(i)   Provide an understanding of the past, present, and potential changes occurring in the subject area of the proposed topic;
(ii)   Provide significant intelligence for strategic decisions;
(iii) Encourage and facilitate strategic thinking; and
(iv)   Assist in the objective formulation.

The introductory section should include the following: background of the study, statement of the problem and or research questions, main objectives, and significance of the study. The *background of the study* is the subsection wherein issues, concerns, or considerations, which informed the proposed study are outlined. The background of the study is presented by utilising logical statements to depict that the proposed research is worthwhile to expend resources to conduct. It is expected to depict the nature, extent, and recent status of concerns influencing the problem, presented in a manner such that it should be clear and convincing to the readership of the proposal.

The *statement of the problem*—the first significant step in writing a research proposal. The problem statement is the foundation for developing a sound and robust research proposal. It facilitates the simplification of the selected topic for research and presents the possibility of a systematic depiction of the research problem. Moreover, the problem statement highlights the importance of the research problem within the selected subject area or in the area of specific research interest as a priority problem, thus emphasising the need for the proposed research. The focus of the study is expected to be specified clearly in scientific terms. Thus, the problem statement should include an overview of the background information on the significance of the socio-economic, environmental, geographical, or political concerns within the subject area or in the area of specific research interest; detailed information on the needs for the proposed research, which have been observed (i.e., established knowledge gaps); the scope of the problem and significance of providing solutions; and the assumed solutions.

The *research questions* are the central questions that a researcher or a team of researchers seeks to answer to with the proposed research. Research questions offer a sound basis to provide descriptive data that could be utilised to depict a better picture of the proposed problem. Simplicity and clarity of language are essential in constructing research questions, which should be short, crisp, and precise. Generally, a probing question that proffers solutions to one's thought or idea is preferred. The research questions and hypotheses may both be included in a single study to perform distinct but complementary purposes. The research proposition that a researcher or a team of researchers wants to verify is known as a *hypothesis*. It is essential to know that while the hypothesis is valuable, it is not always required.

Occasionally, research can be all about gathering and analysis date to identify key characters. Although with problem-oriented research, it is required to formulate hypotheses in clear terms as much as possible. Generally, in such problem-oriented research, hypotheses are concerned with the causes of a specific phenomenon or correlation between variables that are being investigated. There are several essential *steps involved in hypothesis testing*. These include *specifying the null* and *alternative* hypotheses while analysing the research problem; selecting a suitable *statistical test* based on the research design after a decisive sampling distribution, which applies to the selected test statistic; specifying the *significance level* for the research under investigation; gathering data and their computed value of the test statistic that is suitable for the sample distribution; determining the probability of the test statistic under the null hypothesis employing the sample distribution; and comparing the obtained probability with the stated significance level and then rejecting or accepting the null hypothesis based on the comparison.

The *main objective* is to provide a systematic statement of the expected research outcomes. The *significance of the study* (justification or motivation for the study)—statements of why it is necessary to carry out the research or motivation for the proposed topic that are included in the writing of a project proposal. This statement expresses why it is laudable to spend time, resources, and effort, as well as the inherent gains to the society by proffering solutions to the challenges in terms of potential benefits for policy and practice.

The *literature review* section involves a review of theories and archival literature, conducted to assist the researcher to know what the research project is as regards the problem of the study that has been carried out and reported by other researchers. A literature review not only averts reinventing the wheel, i.e., duplication of research work, and allows a researcher or a team of researchers to familiarise themselves with the state of knowledge in the subject area, but it also helps to identify the knowledge gaps in the existing literature. Moreover, the review of the literature presents researchers with insights into the different research methodologies, which have been applied to the study. Following a proposed research methodology, sources of information and data for a thorough literature review of archival literature might be obtained from general references, fieldwork, observation, and published or oral sources. These can be obtained from local, national, and international journals, books, routine statistics and registers, opinions and beliefs from key figures in the communities—locally or internationally—and surveys and annual reports among others. Reports of organisations, for instance, NGOs and international organisations like the DFID, UN, WHO, UNDP, AfDB, World Bank, IMF, UNICEF, and so on are also possible sources of literature.

The *research methodology* section should specify the research methodology, i.e., materials and methods for consideration and or the data types—primary or secondary-to be obtained with a brief explanation of how the materials and methods would be executed and the required research data would be acquired. Based on the nature of the research project, these might be obtained through questionnaires, surveys, simulations, experiments and so on. However, the nature of the research problem will undoubtedly suggest the research techniques be employed. Moreover, it is vital in the section of the proposal to specify how the raw data to be gathered would be refined into researchable data, that is, how data would be sorted or organised, processed, and analysed to research outcomes.

The *expected research outcomes* section should state the research findings that the researcher or a team of researchers, and perhaps the funding body, expects to be accomplished after successful project completion. Though a researcher or a team of researchers cannot predict what the expected outcomes of a research would be, they must endeavour to be quite exact as to the nature and scope of the outcomes and those whom the research findings may benefit [2]. The expected research outcomes must be associated directly with the research aim and objectives. These expected research outcomes might be *qualitative, quantitative,* or *mixed*. However, researchers should endeavour to state the findings quantitatively as much as possible because the funding bodies prudently consider them. The number of expected research outcomes might be three, four, or more, in proportion to the project aim and objectives.

4.2.1.4   Project Timeline

The *project timeline* also referred to as a work-plan is often prepared and included when writing a project proposal because it helps the researcher or a team of researchers to highlight and discuss the project schedule with institutional authorities, funding agencies and supervisors, among others. Generally, a work-plan or work-calendar is a schedule, graph, or chart, which gives an estimate of time for the research project's summary of the various tasks and how they are clearly related. It is imperative to note that a work-plan should be simple, realistic, and easily comprehended by whoever should be reading it. The work-plan addresses not only the research project's preparatory and implementation phases but also the proposed research methodology, i.e., materials and methods, data collection, its processing and analysis; and report writing, dissemination of information, and utilisation of results. Therefore, it typically entails a presentation of the tasks to be carried out, when the tasks would be executed, timeline or required time to perform each task, and who is to perform what task. Table 4.1 presents a simplified project timeline for predominant project activities, summarising research project tasks and their duration in the form of a Gantt chart.

Another terminology associated with a project timeline is the *work schedule*. The *work schedule*, on the other hand, is a table that essentially gives an overview of the tasks to be executed in a research project and the duration of the different tasks. Table 4.2 presents a simplified project work schedule, summarising the tasks to be performed and their duration in a table format.

Table 4.1 Simplified project timeline for an 18-month task

S/N	Description of activity	Duration	Year	Quarter			
				1st	2nd	3rd	4th
1	Background information and statement the of problem	October–November	2020				✓
2	Aim and objectives, significance and scope of the study	December	2020				✓
3	Review of theories	January	2021	✓			
4	Review of relevant published literature	February–March	2021	✓			
5	Configuration of the combined cycle and their thermodynamic processes	April–May	2021		✓		
6	Formulation of governing equations, and modelling and simulation of the cycles	June	2021		✓		
7	Computational and experimental investigations	July–November	2021			✓	✓
8	Results and discussion	December–January	2022	✓			✓
9	Report writing and presentation	February–March	2022	✓			

Table 4.2 Simplified project work schedules for an 18 month task

Task	Duration
Introduction	**3 months**
1. Background information and statement of the problem	8 weeks
2. Aim and objectives, significance and scope of the study	4 weeks
Literature review	**3 months**
1. Review of theories	4 weeks
2. Review of relevant published literature	8 weeks
Methodology	**8 months**
1. Configuration of the combined cycle and their thermodynamic processes	8 weeks
2. Formulation of governing equations, and modelling and simulation of the cycles	4 weeks
3. Computational and experimental investigations	20 weeks
Results and discussion	**2 months** (8 weeks)
Report writing and presentation	**2 months** (8 weeks)

4.2.1.5   Budget

The research project *budget*, also referred to as the total research cost, is an estimate of the required personnel allowances or costs, equipment costs, supplies or consumables including materials costs, combined with expenditures for travels and overhead charges, among others, for a set time. It is an essential element of the research proposal because the research plan is expected to be a strategy that is cost-effective for interested partners. An unjustified, bloated budget is probably going to result in no funding or low-priced project funding. However, an undervalued budget might get a research project funded but result in a probable failure in delivering the stated research proposal's outcomes.

Usually, the budget is not prepared until the research design or planning has ended. Moreover, because the budget cost is generally a limiting factor, the budget should always be kept in mind during the research design phases to prevent writing an unrealistic proposal. For a more financially viable research project, locally sourced materials should be considered. The reasons a detailed, carefully thought out budget is required are [3]:

(i)   It helps a researcher or a team of researchers to identify resources that are readily available locally and others that might be needed;

(ii)   It is a valuable tool for planning, implementating, and monitoring research projects; and

(iii)   The budget design process might assist a researcher or a team of researchers in contemplating some aspects of the project timeline, yet to be considered. Thus, it serves as a valuable reminder of the planned activities as the research project progresses.

Often, for convenience, the budget for a research project is expected to be prepared based on the project timeline or work schedule as the starting point. With a well-defined project timeline, the required budget to carry out the research project can be estimated, which would require *cash* (from donor agencies) and *in-kind* contributions (such as institutional support) [4]. The in-kind contribution, an inventory of resources, is the direct support available for a research project that might not require cash. For instance, the host institution may offer facilities such as lab space, computing facilities, and Internet access. The host institution would, often, attempt to give a cost to these contributions, computed as a specific per cent of the total direct cost. A researcher or a team of researchers should undertake a detailed project design to guarantee that every resource, including personnel, is available.

Typically, industrial-funded projects are managed under different guidelines compared with research grants or projects funded by government or non-profit organisations. These financial provisions are specific to each research project and the research associated reward structure. For instance, when a researcher or a team of researchers might gain materially and economically from the research findings through patents and their commercialisation, etc., they should have a stake in the financial risk. That is, the host institution is expected to provide specific project finances or counterpart funding. If a researcher or a team of researchers does not benefit financially from a funded research project, then the funding body would be responsible for all the financial risk.

The *value proposition* of a research proposal is very imperative [4]. If an organisation is to be approached for funding or not, with a research proposal, it is essential to identify what is significant to the organisation. For instance, the proposal must attempt not just to address one of the organisation's predominant goals but more. This can simply be accomplished by assuming that the researcher or team of researchers made the decisions for funding in response to the research proposal. When a researcher or a team of researchers can persuade a funding body that a proposed project would not only add value to the funding body but also enhance its reputation, in particular, given the spent time and money on supporting the project. Such a proposal stands a chance of being funded. Hence, the proposal should be developed in a manner that reflects the organisation's objectives. Apparently, as an engineer or a team of engineers, proposal writing is also expected to ensure that the award of the research contract would be beneficial to society, which remains a moral obligation.

Table 4.3 presents a typical research project budget, listing specific project items. It listed the standard categories of expenditure and, perhaps, the likely division between funded research expenditure and the in-kind contribution to the project. Several donor agencies cap their cash commitments to specific budgetary items at a certain per cent of the total budget (as a rule-of-thumb). For instance, personnel allowances or costs should not exceed 20% of the budget, costs of equipment should not exceed 25% of the budget, and so on. Moreover, allowances or costs for each member of the research team should not exceed 20% of their monthly earnings, student researchers should be paid 100% of their monthly earnings and each member of the technical support and research assistant payments should not exceed 50% of their monthly earnings. However, when more than one participating organisation is involved in a research proposal, additional in-kind columns are expected to be included to list the in-kind contributions from these other organisations.

Table 4.3 A typical research project budget

Description of items	Project funded	In-kind contribution	Total
1. Personnel allowances/costs			
1.1 Principal researcher	*		*
1.2 Team members × 4	*		*
1.3 Student researchers × 2	*		*
1.4 Technical support × 2	*		*
1.5 Research assistant × 2	*		*
Sub-total			*
2. Equipment (list and specify)		Laboratory Space	
2.1 Expander generator (1 pc)	*		*
2.2 Condenser (1 pc)	*		*
2.3 Feed pump (1 pc)	*		*
2.4 Heat exchanger (2 pcs)	*		*
Sub-total			*
3. Supplies/consumables			
3.1 Computing facilities		Computing Support	
3.2 Stationeries	*		*
3.3 Materials for structural and electrical works	*		*
Sub-total			*
4. Data collection and analysis			
4.1 Research instruments for data collection	*		*
4.1.1 Software for modelling and simulation	*		*
4.1.2 Software for thermo-economic evaluation	*		*
4.2 Instrumentation for data collection			
4.2.1 Digital voltmeter (1 pc)	*		*
4.2.2 Liquid turbine flow meters (3 pcs)	*		*
4.2.3 Pressure transducers (2 pcs)	*		*
Sub-total			*
5. Travels and dissemination			
5.1 Car hire (5 Weeks)	*		*
5.2 Industrial visits and collaboration × 3	*		*
5.3 Flights (Lagos-Berlin) × 3	*		*
5.4 Conference registration × 3	*		*
Sub-total			*
Direct cost			*

(continued)

Table 4.3 (continued)

Description of items	Project funded	In-kind contribution	Total
6. Others/miscellaneous (specify)	*		
6.1 Incidentals (10% of the direct cost)	*		*
Total direct cost			*
Institution overheads (5% of the total direct cost)	*		*
Project cost			*

* The items with an asterisk are expected to be specified budgetary values in an appropriate currency

The *budget justification* for all budgetary items of a research project budget is carried out by a researcher or a team of researchers to justify with logical reasons all budgetary items [4]. The personnel list should have a definite responsibility for all members of the research team involved in a funding application for a research project, which should be included in the budget justification. For all additional personnel who are neither a member of the research team nor named explicitly in the funding application, a brief and clearly expressed description of their qualifications and skills is mandatory. This would allow the funding body to evidently compare the sought funds to the desired skill types, which is expected to be directly associated with the wage level.

Also, the budget justification should match the research proposal's methodology and outlined analysis—a consequential task, requiring considerable thinking and substantial meticulousness [4]. Essentially, the equipment and space that are not itemised in the research project budget but are accessible by the research team or members of the research team, as either part of the budget or somewhere else in the funding application, should be stated. The donor agency may desire to see a signed document by an independent research facility, establishing its readiness and ability to carry out the research work to the required standard. If a cost is involved, it must be listed in the budget. If a researcher or a team of researchers gets funded by a commercial partner, several moral concerns may arise due to *conflict of interest*. Thus, researchers are expected to ensure that presented research outcomes are unbiased and they should clearly state all conflicts of interest in every report, presentation, and publication. This might be done in the paper's acknowledgements section.

4.3 Research Protocol

Research protocol can be described as a project-planning document that follows a standard outline, focusing on the critical thinking of the researcher about the study as anticipated at the beginning of the research [5]. After gaining a better understanding of the research topic and project design, a researcher can commence the research protocol drafting. The supervisor of the researcher should support and provide competent, ongoing supervision to the researcher during the writing of the protocol. A researcher should be mindful of a narrow research focus, a thorough but focused literature search and review of theories and literature, carefully chosen materials and or location, and a cautiously decided (combination of) research methods for the study during research protocol drafting or writing. This process is expected to be a collaborative and supportive process, and no researcher will be successful without the help of their supervisor(s) and or technical staff, as well as administrative assistance from the university or other institutions.

The recommended *phases for a research protocol* are as follows [6]:

Phase 1: Meeting with the supervisor and clarifying the project theme

The prospective researcher should

(i) Visit his/her supervisor to finalise the research project title and design the research project; and
(ii) Draft the protocol and the supervisor render assistance and refers the researcher to the dean of faculty, head of school, and head of department or programme to discuss the research protocol's submission and defense.

Phase 2: Approval of the protocol

After a research protocol defence, the protocol is recommended for approval or otherwise by the department and faculty boards of postgraduate studies or research committee. Afterwards, it is submitted to the university postgraduate school.

Phase 3: Financial assistance through an application for a grant

The researcher can seek funding through an application for a grant from donor agencies.

Phase 4: Supervision and completion of the research project

The researcher executes the research plan while the supervisor provides competent, ongoing supervision.

4.3.1 Goals of Research Protocol

The predominant *goals of a research protocol* are to [6, 7]:

(i) Justify the need for the specific study by convincing stakeholders that there are gaps in knowledge or challenges requiring consideration;
(ii) Suggest that the researcher is competent and prepared to carry out the research as well as manage the research project; and
(iii) Demonstrate that, no matter what the research outputs are and notwithstanding whether the research hypothesis is true, the research will make valuable contributions to knowledge in the field of study.

4.3.2 Contents of a Research Protocol

The following is the preferable sequence for the different *contents* (or *section headings*) *of a research protocol* [6]:

(i)   General information;
(ii)   Project title;
(iii)   Background to the study;
(iv)   Problem statement and or questions;
(v)   Aim and objectives of the research;
(vi)   Significance of the study;
(vii)   Definition of the research scope and limitations;
(viii) Research methodology;
(ix)   Preliminary results (if any);
(x)   Ethical considerations (if required);
(xi)   Chapter outline;
(xii)   Timeline;
(xiii) Budget; and
(xiv)   Bibliography or references.

NB: This sequence of section headings is occasionally separated or combined and may differ from one engineering discipline to the other.

4.3.2.1 General Information

The general information is expected to contain vital biographical details like the *project title, student name* and *number, department, faculty/college/school, institution name, proposed degree* and *date.* All these are written on the cover page. The title page includes the previously mentioned information, as well as some specific salient requirement statements and the name of supervisor(s).

4.3.2.2 Project Title

Typically, the *project title* should be expressed in the least feasible number of words, ten or more, which is expected to provide as much information about the research field and direction as possible. That is, the researcher should identify and state a unique and succinctly descriptive project title that communicates to the readership the protocol, the research field, and direction, thus suggesting the purpose of the study. Also, it must indicate the significant parameters (where applicable) and perhaps should not be in a question form.

4.3.2.3 Background of the Study

The *background of the study* is the significant first step of the research work, providing a summarising review of significant background information on the research title within the context of the subject area.

4.3.2.4 Problem Statement and or Research Questions

Formulation of the problem statement involves the definition of what the research to be conducted proposes to do. The *statement of the problem* should be specific, unambiguous, and presented in a systematic sequence, starting with information or theories necessary for problem comprehension, some justifications that may include citations and a declarative or descriptive statement, or intensification in question form. The research questions, which are the major questions that a researcher or a team of researchers seeks to answer with the research, offer a sound basis to provide descriptive data that could be utilised for depicting a better picture of the investigated problem.

4.3.2.5 Aim and Objectives of the Study

The *aim* is the systematic statement of the expected outcomes of the research, and the objectives of a research project should summarise, in an organised manner, what the study accomplishments would be. The types of objectives are *general objective*—the would-be study accomplishments stated in generic terms; for example, if the challenge identified in bottling manufacturing industries is a high injury rate. A general objective could be to establish system dynamic simulation models for managing the magnitude and risk factors of injuries in manufacturing industries to mitigate against the high injury rate; and *specific objectives*—the would-be study accomplishments, stated in specific terms as interrelated objectives. That is, specific objectives are a breakdown of general objectives into more straightforward, interlinked objectives. For example, from the aforementioned general objective, the specific objectives are to.

(i)   Identify and analyse the common injury risk factors;
(ii)   Determine which department and or sections with the highest injuries rate;
(iii) Determine the age distribution of affected workers;
(iv)   Formulate a mathematical model for determining the magnitude of risk, based on identified factors; and
(v)   Apply the model to the manufacturing industries.

These *aim* and *objectives* must be specific, measurable, achievable, realistic and time-bound, i.e., the acronym *'SMART'*.

4.3.2.6 Significance of the Study

The *significance of the study*, also known as the justification or motivation for the research, provides a clear description of the reasons the study is required, that is, the rationale for the research.

4.3.2.7 Research Scope

The *research scope*, also known as the scope of the study, states the contents and extent of coverage for the study that the research would address with required and available resources. It involves the study delimitation—specifying the limits or boundaries of the study to be covered. Also, the scope of the study explains in detail the research limitations, i.e., the extent of the research studies—what would and would not be investigated. Far from being a hedge, spelling out these concerns at an early stage in the research might prevent misunderstandings and disagreements on conclusions drawn, after the completion of the project.

4.3.2.8 Research Methodology

The research methodology is a set of scientific actions, techniques, and instruments utilised in carrying out a research project by proffering solutions to the research problem, answers to the research questions, and achieving the specified objectives. It specifies the materials and particular methods that should be observed to undertake the project research. Essentially, a researcher should study the different methodologies that can be employed to carry out a proposed research. The decisions on the proposed research methodology can only be taken after a researcher has done both an initial literature search and review and a rigorous examination of all relevant studies and texts on possible research methodology.

4.4 Design for Research Outcomes

Generally, a research plan is designed to provide conclusive solutions and answers to research problems and questions, respectively, which should include the analysis and interpretation support for the results. With a multi-parameter investigation, consisting of controlled or uncontrolled parameters, one or more parameters (i.e. the dependent variables) can be measured as a response due to changes in the independent variable (i.e. another parameter) based on the nature of the investigation [4]. The consistency or uniformity of the measured data is a measure of the degree of their strength, which might be observed by examining the set of scattered data points around the trend line. However, the measurement of these scattered data points is of great value or significance for establishing the trend line accuracy.

Nearly every published article in engineering employs figures as a block diagram to present the experimental or survey method and summarise findings utilising charts, graphs, etc. While developing a research plan, a researcher or a team of researchers should cautiously decide on how the findings might be presented in a manner that is typically unambiguous or not open to more than one interpretation. Often, skimming readership through the article should be enticed by the self-explanatory graphical representations, illustrating the research findings. They must be capable of gaining a level of understanding from the visual representations and their accompanying legends, exclusive of the detailed explanations given in the main body of the article. Consequently, a researcher or a team of researchers writing a research proposal is expected to dwell on what the readership of the proposal could gain from the appreciation of the research outcomes presentation and which methods of data analysis and their interpretation can be utilised to justify the research conclusions.

4.5   Research Tools and Techniques

One of the numerous challenges, in particular, with all experimental engineering research, is how to check or prove the validity of the research findings by employing independent means. The research plan should briefly state the methods to be utilised for carrying out such result validation. The most important check or validation of the experimental research findings can be undertaken by comparing the findings against the findings of a theoretical model. If two distinct and independent experimental research techniques generate results that are compared favourably statistically, the employed tools for both research are highly valid. However, there is a likelihood that, due to the same bias, both techniques might yield similar research outcomes, meaning that both the obtained research findings and those from the published literature might probably be incorrect. This happens when similar mistakes, common to both methods of analysis, are committed by omission or commission.

4.5.1   Experimental Measurement Tools

In general, a standard experimental measurement tool offers the best and most reliable means of obtaining consistent measurements. When alternatives are employed, a calibration procedure should be used by the researcher or a team of researchers to calibrate the measuring tools or verify that they are appropriately calibrated [4]. With nearly all experimental setups that require alternative measuring tools, the procedures for calibration of the tools are often well-defined, which involve employing established standards, properties, measurements, etc.; for example, standard solutions for chemical analyses, standard temperature for heat transfer measurements, standard loads for measurement of microwave and electric power levels, application of materials with well-known properties, and the 'null' calibration methods—running the tool on blank samples such as air or distilled water. Furthermore, researchers can conduct the calibration of tools by obtaining measurements of materials and systems from previously published findings in the verified refereed literature.

4.5.2 Numerical Modelling and Simulation

A researcher or a team of researchers can employ modelling and simulation techniques to gather information and data. They can gather the information and data through *numerical* or *mathematical modelling*—the development of a model, based on theory and observation for predicting its performance on mathematical equations; and physical modelling—the development of a physical model for a whole system in a scaled-down version to observe its performance at different conditions [6]. In contemporary times, there are a plethora of computing (numerical or computational) codes for the modelling and simulation of different physical effects [4, 8]. For instance, researchers can routinely employ codes in computational fluid dynamics, engineering equation solver (EES), electromagnetics, and so on for predicting or validating experimental research findings. The computing codes can be either differential, integral, or mixed (integrodifferential) codes.

With the differential codes, the object's surrounding media is expected to be incorporated into the model, and the solution space's boundaries should be at an appropriately sufficient distance to the object such that either the area or volume of interest is not affected, which significantly influences accuracy. The integral codes do not require the boundary definition of the solution space. Problems over a set of physical equations, called *multi-physics* challenges, involve different discretised sets of integral and differential equations, which are integrated into solving the challenges, covering more than one numerical or computational discipline [4]. When computing software is employed as a method of analysis for experimental research, it is required that the models be simulated through the computation of solutions to the standard models' experimentation, and the researcher can compare their findings with theoretical calculations.

Computing software can detect mistakes that are so challenging to detect by employing the models' images or codes. Typically, the errors comprise the imperfect continuity among the materials and their boundaries as well as material properties. A material boundary is defined as the distance between two adjacent nodes, where the distance between the nodes is the spatial resolution limit. Also, there are discretisation limits, i.e., If the model segments are too small or large, the computing program might encounter rounding errors, or the accurate continuous media representation is lost, respectively. However, it is an expected or established procedure to change the model's mesh size for rerun computations where solutions are very alike; hence, the solution is highly likely to be valid. These solutions must also compare favourably with the established analytic solutions (in textbooks if available).

4.5.3   Theoretical Derivations and Computations

With the development of a mathematical case on the basis of its theoretical or conceptual framework, the validation of the solutions becomes obligatory, which can be
accomplished using a number of methods. These methods include the examination
of where the possible solution lies within the anticipated physical boundaries; evaluation of theoretical findings' units through dimensional analysis; and investigation
of findings' validity through mathematical programs (such as Matlab and EES that
offer symbolic solution techniques for integrodifferential equations). Other methods
include the substitution of standard values into formulated equations to ascertain
or make sure that the findings are realistic, and making negligible modifications to
the model input parameters to ascertain if the changes in the simulation results are
insignificant [4]. In the development of a research plan, a researcher or a team of
researchers needs to state the techniques for the validation of the proposed computing
codes as a procedure for the verification of the research findings.

4.5.4   Curve Matching

There are numerous techniques to prove that a set of measured experimental data
tends towards a specific scientific, mathematical equation. One of such techniques is
the use of the equation's independent variables for computing the dependent variables
and their plot against the measured dependent variable, which is expected to be
a straight line for the equation that is true or the data is valid. The plot for the
analysis of the linear regression would provide the findings' statistical significance,
and the line visual inspection would indicate the level of unsystematic or random
mistake and the ranges of the data where the relation fails. It is unfortunate that
systematic or non-random mistakes may cause this method of analysis to lead to an
inconclusive conclusion or result. These systematic recurring mistakes may comprise
the likelihood of an offset angle in a trigonometric function (like the sinusoidal
dependency). Thus, it is imperative to evaluate this offset value such that the curve
may be matched with the possible least mistakes.

4.6   Chapter Summary

A well-developed research plan should communicate an overview of the significant
background information; the nature of the research problem to be solved and or
research questions to be answered; necessities or methods required to obtain the
data for problem solving, data analysis and interpretation of findings; and the techniques for validating research outcomes. Also, it should communicate the required
project timeline, funding to carry out the research project, the manner of reporting

the research findings to the stakeholders plus the scientific communities, and the expertise of the researcher or team of researchers.

Research proposal, a document detailing the description of the intended research plan or programme for consideration, is a framework of the research process that provides the readership of the proposal with an overview of the research project to be conducted. Some of the reasons research proposals are written include requesting a research grant, ethical certification, or as a task in tertiary education. An extensive but focused literature review; a methodology section far more comprehensive than that of a scientific article; a section describing the expected research outcomes; preliminary result(s) (if any); a realistic project timeline; a justifiable budget; and references should be included in a research proposal.

Research protocol, a project-planning document, is focused on the critical thinking of the researcher about the study. When writing a research protocol, a researcher should keep in mind a narrow research focus, a thorough but focused literature search and review, and carefully chosen materials and a cautiously decided (combination of) research methods for the study. The preferable sequence for the different section headings of a research protocol includes the background of the study, problem statement and/or research questions, aim and objectives of the study, significance of the study, the definition of the research scope and limitations, literature review, research methodology, preliminary results (if any), ethical considerations (if required), chapter outline, project timeline, budget and references.

Research findings should be validated by employing two separate research tools and techniques or more. To validate a research technique, a researcher or a team of researchers may utilise the boundary cases and simple system configurations. At the research design stage, a researcher or a team of researchers must briefly state how they would present the research findings. The presentation method should be brief and clearly expressed in a logical manner such that the final report's readership would be able to draw similar inferences as the researchers. For fund seeking and other support, the research proposal is expected to clearly state how the researchers would communicate the research project, and its findings, and the benefits of the research findings to the donor agency as well as the public at large.

References

1. 5StarEssays. (2020). Writing a research proposal—Outline, format and examples. In *Complete guide to writing a research paper*. Retrieved from https://www.5staressays.com/blog/writing-research-proposal
2. Walliman, N. (2011). *Research methods: The basics*. Routledge—Taylor and Francis Group.
3. Olujide, J. O. (2004). Writing a research proposal. In H. A. Saliu & J. O. Oyebanji (Eds.), *A guide on research proposal and report writing* (Ch. 7, pp. 67–79). Faculty of Business and Social Sciences, Unilorin.
4. Thiel, D. V. (2014). *Research methods for engineers*. University Printing House, University of Cambridge.
5. Mouton, J. (2001). *How to succeed in your master's and doctoral studies*. Van Schaik.
6. Lues, L., & Lategan, L. O. K. (2006). *RE: Search ABC* (1st ed.). Sun Press.

7. Bak, N. (2004). *Completing your thesis: A practical guide.* Van Schaik.
8. Sadiku, M. N. O. (2000). *Numerical techniques in electromagnetics.* CRC Press.

Chapter 5
Project Report Writing and Presentations

Abstract The objectives of this chapter are to

(i) Specify and discuss the basic chapter titles and sub-titles in a research project report;
(ii) Prepare a research project report based on the basic format;
(iii) Specify and explain the basic methods of writing bibliography, citing and listing references;
(iv) Discuss research presentations;
(v) Describe plagiarism and citation; and
(vi) Discuss citation and its management.

Keywords Project report · Project presentation · Conference presentation · Plagiarism · Citation · Referencing style · Citation management

5.1 Introduction

Project reports—dissertations, theses, and technical reports, which are typically written in chapters, sections, and subsections, have three sub-divisions—preliminary pages, the main text, and supplementary pages. The *project report writing* is the final stage of any research and communication process where its predominant objective is to *present or disseminate the research project and its findings* to stakeholders and the scientific communities. The preferred mode of project report writing is the *impersonal mode,* i.e., instead of *'I conducted the study to investigate ...'* or *'The researcher conducts the study to investigate ...'*, say, *'The study was conducted to investigate ...'*.

Like the research proposal, the main text of project reports is identical to the first three sections of research proposals except that research proposals are written in present and future tenses, while those of the project reports are in reported speech. Except for the use of tense, the proposal becomes the initial contents of a project report, i.e., the first three chapters, when written in detail. In presenting a project report, a researcher or a team of researchers should follow the required in-house style or format as preferred by most institutions or publishing outlets. These in-house styles or formats do not vary significantly from the standard format.

5.2 Writing a Project Report

A *project report* can be described as a document that details the activities carried out during the process of conducting a research project and its findings by a researcher or a team of researchers. Project report writing is one of the final requirements of any research and communication process, and the presentation of the research outcomes and their publication are essential for the dissemination of the research findings to peers within the scientific community [1]. The project reports are typically presented in a specific standard format, which includes *preliminary pages*; the *main text* of the project; and *supplementary pages*.

5.2.1 Preliminary Pages

The *preliminary section* comprises the title page, declaration page, certification or approval page, dedication, acknowledgements, abstract, table of contents, and list of tables and figures. The preliminary pages are numbered with Roman numbers in lower case, i.e., i, ii, iii, and so on.

5.2.1.1 Title Page

The *title page* is one of the preliminary pages and the first page in the section, comprising the *project title*, *author's name*, the research relationship to a degree, diploma, or course requirement, the institution's name to which the report is to be submitted, and the month of presentation. The *project title* must be unique and succinctly descriptive, which should communicate to the readership of the project report the research field and direction, suggesting the purpose of the study. One of the basic elements that should be contained in the project title is the key project parameters (where applicable). It should be typed in *single-spaced*, *bold*, *capital letters*, and *centre-justified* between the right and left page margins.

5.2.1.2 Declaration Page

The *declaration page* is a confirmation page that the work was actually carried out by the researcher and not copied from someone else's work. This page should be signed and dated by the researcher. A sample of the declaratory statement is as follows: '*I, name of the researcher (with registration/matriculation number), hereby declare that*

this project report or thesis entitled: "Model Development and Experimental Valida-
tion of a Solar-Biomass Hybrid Power Generator System" is a record of my research
project. It has neither been presented nor accepted in any previous application for a
higher degree. All sources of information have been specifically acknowledged'.

5.2.1.3 Certification Page

The *certification or approval page* includes the originality attestation for the project,
and some of this information are as follows: the names and registration (matricula-
tion) number of the student; project title; and the research relationship to a degree,
diploma, or course requirement; as well as the names, signatures of the head of
department, supervisor(s), external examiner and dates, which varies from institution to institution. A sample of the statement for certification is as follows: *This is*
to certify that this project report or thesis presented by 'name of the researcher'
(with registration/matriculation number) entitled: "Model Development and Exper-
imental Validation of a Solar-Biomass Hybrid Power Generator System" has been
read and approved as meeting the requirements of the Division of Mechatronic and
the 'name of the University' for the award of a Masters of Engineering (MEng)
degree in Mechatronic Engineering.

5.2.1.4 Dedication

The *dedication page* is optional in a project report, where emotionally laden words are
permissible to honour people whom the researcher cherished or who have contributed
in a manner or the other to the successful completion of the research project, and those
who, in particular, have shown curiosity or concern about the research outcomes.

5.2.1.5 Acknowledgements

The *acknowledgements* are restraining and straightforward statements that are used
to express appreciation to individuals and or organisation(s) who have assisted during
the process of research proposal writing and doing the research project and its report
writing—drafting, editing, and final editing.

5.2.1.6 Abstract

The *abstract* is a concise synopsis of the research project, comprising a brief *back-*
ground to the study; the *main objective* of the study; and succinctly described *mate-*
rials and methods for data collection and their *analysis and interpretation*; *research*
findings; meaningful *conclusions;* and the *implications of the study.*

5.2.1.7 Table of Contents

The *table of contents* presents in sequence the contents of a report outline, laid out in a tabular form. From the preliminary pages through the pages of the project's main text to the supplementary pages, all the report's chapters, headings, and sub-headings with their corresponding page numbers are serially listed.

For presenting tables and figures in a project report, a separate page is utilised for the *list of tables* and *list of figures*. Such a page exhibits sequentially listed, numbered tables or figures with their corresponding titles and respective page numbers in the report. Also included in the list are appendices' numbered tables or figures with their corresponding titles and respective page numbers, which are embodied in the report or annexed to it.

5.2.2 Main Text of the Project

The standard format for the *main text* of a project should comprise five chapters: *introduction*, *literature review*, *methodology*, *results* and *discussion*, and *conclusions* and *recommendations*, with well-established sections and subsections in each chapter. The readership of these project reports would look out for these chapters and their sections and subsections. As a result, it is recommended that researchers should not deviate from this standard format unless where the research sponsor, examining body, or publishing outlet, as the case may be, specifically stated that they should do so or can do so.

The *introduction* consists of the background to the study, statement of the problem, research questions and or hypotheses (if applicable), aim and objectives, significance of the study, research scope and definitions of terms. In the *background of the study*, issues, concerns, or occurrences, which informed the investigations are outlined. The background to the study entails the use of logical statements to depict that the proposed research is worthy of expending resources on doing the research. The *statement of the problem* is the definition of what the research to be conducted (by a researcher or a team of researchers) proposes to do, showing the existing gap in knowledge that the research intends to fill. The problem statement should be succinct, well-defined, and convincing information on the topic and planned parameters for investigation. It must be specific, a matter of fact, and presented in a logical sequence, starting with information and theories necessary for problem comprehension, some justifications that may include citations, and a declarative or descriptive statement or intensification in question form.

The *aim and objectives of the study* should state the particular viewpoints of a study and the rationale behind the focus on them, including declarations of what would be achieved and investigations that would be considered. The *significance of the study*, often called justification or motivation for the study, is a set of reasons or a logical basis for the study, which elucidates the usefulness, utility value, or relevance of the study as well as the individuals or groups, government or private institutions that would gain or benefit, and how such gains or benefits might be invaluable. Also, the significance of the study enlightens on how the research findings would contribute to advancing the knowledge frontiers, raising novel questions or disclosing knowledge gaps that the study would fill. The *scope of the study* states the contents and extent of coverage that the research project would address.

The *literature review* entails a review of related conceptual frameworks, theories, and literature. The *research methodology* entails the materials and methods, data collection and methods of data analysis. The *results and discussion* involve data analysis and the interpretation of findings and their discussion, and an overview of the significant findings. The *conclusions and recommendations* include the significant inferences of the study, the limitations of the study, and suggestions for future work, respectively.

5.2.3 Supplementary Pages

The supplementary pages include the bibliography/references, appendices and the index.

5.3 Research Project Presentations

Research project in engineering is usually reported, in written form or orally, using the *hourglass structural model* [1]. Like the hourglass structural outline, research project reporting starts with a broad spectrum of *background* to the study and *review of literature* for the desired information to a narrow description of the *research methodology and presentation of results* (like the neck of an hourglass). Afterwards, the research is concluded with another broad spectrum of *discussion of results* and *conclusions* to expatiate on the implication of the results on the existing body of knowledge. Researchers should utilise this approach for any scientific and engineering project research and communication process, including the undergraduate, graduate, and industry reports of workshop and lab experiments, but not patents. Figure 5.1 depicts the hourglass structural outline for scientific and engineering project research and communication process, indicating the two broad spectrums and the narrow neck. It illustrated the standard format that a researcher or a team of researchers should follow for communicating their project and presentation, conference or journal paper presentation and publication, and so on.

First and foremost, the beginning part of the introductory section for any engineering paper manuscript to be submitted for presentation and publication must typically address both the expert and non-expert audience. That is, the research is expected to be committed to a far-reaching context within the subject area, addressing the research relevance and the question of '*how would the research contributions be beneficial to humanity*' [1]. The manuscript's concluding section should similarly

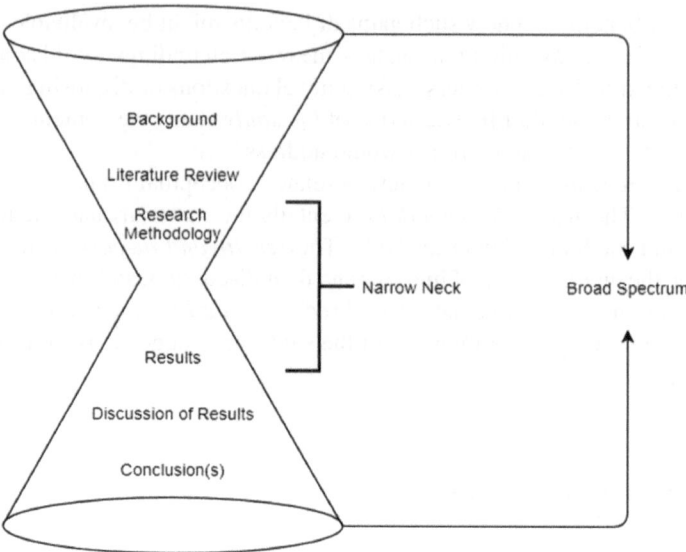

Fig. 5.1 The hourglass structural outline for scientific and engineering project research and communication process. *Source* (Thiel [1])

address a wide-ranging humane context within the subject area, emphasising the main findings and answering the question of *'how a researcher could apply research outcomes and their innovative implications to the benefit of humanity'* [1].

In between the introductory and concluding sections of the paper manuscript, the researcher or a team of researchers is expected to describe the methods to be employed for research data collection and their analysis and interpretation. On the basis of the analysed data and their interpretations, concluding inferences are drawn. Thus, a researcher or a team of researchers should provide a precise *research description* and the *primary data* such that listeners during the paper presentation or the readership of the conference paper after publication would have adequate information to reproduce similar conclusions [1]. With full-length papers, a researcher or a team of researchers in the engineering discipline must be capable of reading the paper to get a piece of comprehensive knowledge of the *materials* and *methods* for research data collection, such that other researchers can independently undertake the same procedures for research data collection, generate similar data, and draw similar conclusions.

Researchers seeking to present at any engineering science conference and want their conference paper published should thus get involved in the process of review in a positive light and respond to reviewers' criticisms of the submitted manuscript in a cautious manner. The method of peer review not only enhances the merit case of the paper by bolstering its background literature, novelty and arguments, among others, but also enriches its logic. Understandably, it can be disappointing and disheartening for a researcher or a team of researchers to have their manuscript rejected. However, the reviewers' predominant rationale is to guarantee that a presented and published paper is comprehensible in its stated contexts and that it is logical and valid in every way in which the stated contexts could be interpreted or understood. Reviewers' comments on submitted paper manuscripts, for all researchers, in particular first-timers or beginners, constitute a free experts opinion.

Though a researcher or a team of researchers has not only the comparative advantage of gaining renowned scholars' attention in the engineering field when their paper is remarkable and sound in reasoning or with convincing evidence, it can simply be advantageous to the reputation of the researcher or research team, and submitting a low-quality research and a deficiently written manuscript would harbour doubts about the researcher or research team's reputation. For the reasons aforementioned and for the scientific community's public good, research outcomes reporting demands substantial time for clarity of purpose and primarily, numerous revisions are undertaken before submission of a prepared manuscript to a conference technical programme committee or the editor-in-chief of a journal editorial board [1]. Similarly, and most importantly, researchers should undertake these before submission of a prepared project report to a project supervisor, examiner, or lab manager in academia or industry.

However, the experience of communicating research outcomes to an audience is not enjoyable for numerous individuals. Still, research presentation is imperative in disseminating new knowledge to the scientific community, which also makes the name, skills, and expertise of the researcher or research team known to them. Unlike project reports, journal articles, or other scientific papers; a research presentation presents an avenue to communicate ideas to a captive audience and get feedback from peers, which may influence the present work or contribute to future work. The predominant goal of any research project presentation, be it a conference, poster, seminar, workshop, or project defence presentation, is to communicate ideas and their importance [2], and get feedback.

5.3.1 Patents

Patent, an authority or licence by the government of a sovereign state, is granted on the basis of the originality of an idea, innovation, invention, or research findings. So, when there is a necessity for a researcher or a team of researchers to protect the intellectual property of their idea or research findings, they must file a patent application way ahead of any conference or journal paper submissions. Often, patent writing is carried out by hired experts due to the technicality and legal framework of the process. However, researchers can do it themselves by learning the nitty-gritty. A researcher or a team of researchers can file patents for an idea, innovation, or invention without scientific verification because it is simply the concept, as well-defined by the *claims*, that is significant [1].

In patent writing, the *list of claims* is the predominant element. These claims, a series of brief and clearly described technical advances on an idea or research findings, constitute the evaluation basis of the patent [1]. Therefore, the claims should provide adequate information and qualifying statements to assure their uniqueness. Moreover, there is a considerable emphasis on making references to any published related patents, together with the published literature on the topic within the subject area. This might be challenging for researchers with little or no knowledge of patent writing and searching. Basically, a patent is granted based on whether the *idea, innovation, invention,* or *research findings* are *novel,* or *no expert* in the field who can deduce the idea, innovation, invention, or research findings (from the knowledge of published patents and the literature).

5.3.2   Conference Presentations

Submitting research outcomes to a conference for presentation and presenting the accepted conference paper at the conference is a significant way to engage with peers in the scientific community [1]. All engineering science conferences would have a *technical programme committee* in charge of assessing submitted papers for presentation and publication, which would take 3 to 6 months or more ahead of the conference date. Typically, the hourglass structural outline for scientific and engineering project research and communication process is the standard format of a conference paper for submission and presentation.

In most engineering science conferences, a unique and succinctly descriptive *project title* that communicates the research field and direction to the listeners or readership of the conference paper, and an *abstract* of about 150–300 words that provides a complete synopsis of the research project, and keywords of five to seven words are requested by the technical programme committee for assessment and presentation at the conference. Sometimes, a full conference paper is requested; that is, the researcher or a team of researchers should provide a precise *research description* and the *primary data* such that listeners during the paper presentation or readership after publication would have adequate information to reproduce similar conclusions. Whichever way, a full conference paper would be submitted after a review process, if the abstract or the paper is accepted. For an accepted conference paper, its copyright assignment should be scrutinised to ascertain that further publication is allowed.

However, the researcher must paraphrase the wording of the conference paper and its contents enhanced before submission as a manuscript for publication in a journal publishing outlet to eliminate the possibility of self-plagiarism and consequent manuscript rejection.

5.3.2.1   Preparation of Conference Presentations

Typically, a computer set of images referred to as '*slides*', which can either be a *Microsoft PowerPoint* or an *Adobe file*, would be required by the conference technical programme committee from the presenter [1]. The slides would be projected with the aid of a digital projector at the conference during the presentation. The

Microsoft PowerPoint and Adobe files are the most widely used files for presentations at conferences. Every conference is planned and structured to run with precise allotted time limits for the presenters to deliver their presentations. All conference sessions are presided over by a *chairperson* who is in charge of a session's timely presentation of allotted papers and a *rapporteur* to report on the session's proceedings. All presenters get a *start, presentation end,* and *question time slot,* which must be known ahead of the conference, and should strictly be adhered to by the presenters and the chair of the session.

The presentation structure should follow the hourglass structural outline of the engineering project research and communication process. The specific standard presentation tip that could be employed for estimating the required presentation time is to present one slide per minute. Based on this and the number of presentation slides, presenters can estimate the required minimum time to present their slides. Some general presentation tips are as follows [1, 3–6]:

(i)   A presenter is expected to present a slide per about a minute because any presentation rapid than that may be hard for the listeners to follow;

(ii) All tables and figures (including graphs) should have numbers with their corresponding titles, serially numbered. Also, equations must be serially numbered and all slides appropriately titled, which would permit any member of the audience to note and signify the appropriate slides for deliberation during question and answer (Q&A) time;

(iii) The word count of text on any slide must be fewer than 50 and should not be fragmented into more than seven bullet points;

(iv) The references on the presentation slides should be cited (if required) as name and date only and listed after the main text of the concluding presentation slide;

(v)   A presenter should select a plain white or light neutral-colour background with visibly contrasting texts, tables, figures, and equations for clarity;

(vi) A presenter should cherry-pick the font type and its size for the best clarity, which typically means no font size of less than 20 points; and

(vii) At the conference venue, a presenter should do a quick check of animated images on the computer and the setup display facility to guide against any compatibility challenges associated with utilising a computer and a projection device.

5.3.2.2   Delivery of Conference Presentations

A day or more ahead of the presentation, the presenter should do the presentation rehearsals aloud to a number of listeners or alone for ensuring that every abstract idea can be expatiated in a brief and clearly expressed manner. Also, this offers the presenter the possibility of phrasing sentences and ensures that s/he could communicate the presentation within the possible allotted time. There are specific, simple approaches to eliminate nervousness before or during the appearance of a presenter before an audience. These are as follows [1]:

(i)   Most importantly, practise the presentation for a few times ahead of the event, both alone and with a small group of audience;
(ii)   Commit to memory the first two or more statements that would be articulated, allowing the presenter to begin the presentation with little or absence of any hesitation;
(iii) Ensure that on all slides, a presenter must have something more than just the text on the screen to expatiate on because presentations made by reading the screen can only be typically dull to the audience; and
(iv)   Do not start presenting until the title slide is projected, and the session chair has introduced you.

Typically, a conference is organised with a schedule that permits presenters in a session to submit their presentations online or load them from a USB (thumb) drive or memory stick at the conference venue ahead of the commencement of the paper presentation sessions. When presenting, the presenter should face the audience and make an effort to speak audibly in simple statements that expatiate more than just the screen-written text. To present a table, the table's *number and its title*, that is, 'Table 7 presents the variation effects of temperature changes with humidity', should be stated ahead of making any comment on the presented table's data. To present a graph, the graph's *figure number and its title*; that is, 'Fig. 11 depicts the variation effects of temperature changes with humidity', should be stated, and the researcher should also state the axes ahead of making any comment on the presented graph's data. When presenting an equation, the researchers should state the equation's number, explaining both sides and defining key symbols orally.

Essentially, the presentation slides should have a slide that clearly define the end of the presentation, which is simply achieved by having a slide with the words *thank you* or *thank you for your audience* only, and maybe with a *thank you* or *thank you handshake image* for colourfulness. Then, the presenter ought to thank the listeners for their rapt attention to signal the end of the presentation, so that the session chair, if time allows, can initiate the Q&A time. The presenter should never request for questions because it is primarily the role of the session chair. Also, presenters should not exceed their presentation time limit. For a presenter to exceed the allotted time is awfully inappropriate for the session chair (who is responsible for time management), the subsequent presenters (that their presentation is not less important) and the audience (who might want to swap sessions for another to listen to a paper presentation, scheduled to start at the end of the present presentation, or want to go for a coffee or lunch break, among others). Often, a conference has two or more sessions running pari passu in separate rooms. In contrast, a large international conference may have numerous parallel sessions running pari passu, such that a session swap for another presentation room can take a while.

5.3.3 Poster Presentations

At most engineering conferences, the poster sessions provide an author or a team of authors with a significant advantage of opportunities to engage in the event with enthusiasm and present their research project one-to-one to interested parties. In contrast to a conference presentation, poster sessions permit lengthier deliberations of the reported research project's merits and qualities with unlimited time for questions, which is one of the predominant reasons an author or a team of authors would choose to present their research project at a poster session. Additional merits comprise the opportunity to deliberate on the project and its findings without stress due to the challenges of nervousness in making a presentation before an audience and responding to questions. An author or a team of authors at poster sessions could ask questions from the audience, which is unlike the usual practise during the Q&A time after a conference paper presentation.

The author or a team of authors are required to prepare a large poster in either an AO or A1 format with a sufficiently large project title, text font size, diagrams (including figures), and tables that are readable at a distance for a poster session [1]. The author uses the poster during the poster session, by standing close by it, to elucidate on the reported research project, respond to questions on the research project, make comments, suggestions, and observations on the present and future studies, etc. Because the author may not be able to, at all times, attend to the poster or speak with every interested party, or passers-by, posters for poster session must reasonably be self-explanatory. For a poster session, the authors mount their posters way ahead of the beginning of the session and should be close to their posters to discuss them during the planned viewing sessions, which often can be for a day or more. Essentially, the needed ingredients for an author or a team of authors to make a fruitful engagement during a poster session include excitement about the project and good knowledge of the project's basic theory. For an author to attain the maximum opportunity at a poster session, they must be able to offer *an outline of the project* and the *predominant research findings* in about three to five minutes, and respond to questions. Also, the author can provide further vital information, more generally, about the research such as relevant papers that have been published on the topic, and facilitate the exchange of contact details to foster further discussion during the conference and, perhaps, after it into the future.

5.3.4 Paper Preparation and Review

Typically, paper manuscript writing for scientific publication should comply with the standard section headings as follows: *introduction* (including background/literature review), *methodology, results and discussion, conclusions, acknowledgements,* and *references.* However, the usage of specific headings rather than the listed standard section headings is more desirable because it eliminates the necessity for introductory statements at the start of each section [1]. Often, there are *page limits* for most science and engineering publications, which authors should strictly follow. Furthermore, manuscripts exceeding these limits might be rejected automatically, or the over-limit pages would attract substantial page charges.

Most journal outlets welcome complete paper manuscripts that are not up to the total page limit, even when it is considerably less, because there are no benefits in having excessive, repetitive, or irrelevant information on a paper. This page limit is applicable to the total contents of the paper manuscript, which includes the *manuscript's title*, its *abstract*, *main text* (with all tables, figures, and their captions), and *references*. Essentially, all tables, figures, and equations should follow these guidelines [1]:

(i)   All tables and figures should be assigned a number and a title, and be referred in the main text of the paper;
(ii) All tables and figures must have self-explanatory captions that do not require the readership of the paper to read the main text to ascertain their importance and significance;
(iii) A presenter should differentiate the lines of a graph (i.e., a graph with more than one line on the same plot) by employing a legend, label, or other tags;
(iv) All equations should have a number, and their symbols should be well-defined with the closest text to the equation where a symbol is first stated; and
(v)   Every symbol should be distinctive, i.e., all symbols should be unique and be utilised only to denote one parameter or quantity.

Many journal publishing outlets demand that authors submit their paper manuscripts by utilising a template. This is termed '*camera ready*', meaning that authors should submit their manuscripts in the standard style of the journal publication outlet at the first submission [1]. Consequently, an author or a team of authors must endeavour to meet the requirements of the stipulated style. Other journal publishing outlets request that authors submit the paper manuscript's title, abstract, keywords, and authors' names and affiliations on a web page, and the main body of the paper, all tables, and figures as separate files. Once these sectional submissions have been confirmed, the manuscript is compiled into a single Adobe (*.pdf) file. Afterwards, the authors must review and check this file before making the final online submission to the editor-in-chief of the journal editorial board.

The review process is initiated by the editor-in-chief, who would allocate an associate editor to nominate and engage the services of two experts or more in the field (of the paper manuscript) as reviewers. A couple of journals require that manuscripts' authors suggest names and contact details of qualified experts—three or more who can review their paper. Immediately, the contacted reviewers accept their nomination; the associate editor sends the manuscript file for review by the reviewers. For the process of a blind peer review, the authors' names and affiliations are excluded from the sent manuscript file to the reviewers to eliminate any *possible prejudicial effect of nationality, lab prestige,* and so on [1]. The reviewers are expected to make remarks on the *reported technical contents* and the *quality of the language* of the manuscript, and to recommend if the manuscript can either be *accepted* (without any required changes, with minor editorial changes, or with major changes), or *rejected* (typically due to inadequate or absence of substantial technical advances in the manuscript, or the manuscript's parts have been plagiarised (i.e. directly copied), or are technically inadequate, incorrect, or irrelevant). The reviewer should justify this decision with some general observations on the manuscript.

A review of all the reviewers' reports is conducted by the associate editor to decide on the manuscript. The editor-in-chief would communicate the associate editor's decision and the reviewers' comments to the authors for their information. If the manuscript is accepted (with minor editorial or major changes), a revision of the manuscript is required. After the revision, authors must submit a *revised manuscript*, and *their response to reviewers' comments*, detailing the changes that have been made to the manuscript. Often, the revised manuscript and the authors' response to reviewers' comments might be returned to the reviewers for a repeat of the review process until the associate editor makes a final decision of acceptance or rejection of the manuscript. This process of review might be conducted once, twice, or more times, based on the associate editor's decision and the reviewers' comments. Typically, each reviewer involved in the peer review process of a manuscript should agree that it is acceptable for publication before the associate editor would then recommend the manuscript for publication.

5.4 Plagiarism and Citations

Whenever a researcher or a team of researchers utilises any idea from referenced publications as a source for their opinions or extracts from published works in the literature to buttress their points, discussions, or arguments, the researchers should acknowledge these sources with the respective *citations* to avoid *plagiarism*. These include verbatim quotations, paraphrased statements, copied tables and figures, among others.

5.4.1 Plagiarism

The term '*plagiarism*' can be referred to as the presentation of someone else's ideas or works as one's ideas or works in paraphrased words or not, either with their permission or not, by representing them in personal writings without acknowledgement [7, 8]. It is unethical to present the ideas or works of others, even when paraphrased into personal words, without acknowledgment by citing them as a source [9]. Plagiarism can be *intentional* (or reckless) and *unintentional*. Under the work ethic and examination principles, plagiarism is a disciplinary offence, whether it is unintentional or not, because it is a breach of scholarly integrity and a matter of grave concern (i.e., giving cause for alarm). However, the pressure to write project reports or publish without plagiarising is one of the most significant challenges in the changing world of scholarly research. It influences practically every aspect of *research, report writing, and publishing processes*, and this pressure is rising in its significance.

Moreover, globalisation and technology are revolutionising research, and these radical changes impact researchers, institutions, and publications across the globe. Due to these global and technological influences on the world of scholarly research, the modes and spurs for scholarly plagiarism are on the rise. There are numerous reasons to avoid plagiarism. Plagiarism must be avoided by researchers because it is an unethical academic fraud and a breach of scholarly integrity, which undermines the ethos of academic scholarship, institution standards, and the degrees it awards, as well as because researchers, institutions, or publishers aspire to write and publish an original, high-quality report. Thus, it is of vital significance to fully comprehend the *use of information sources and citation principles* due to both the ease it provides in avoiding plagiarism and the improvement benefits it provides in the clarity and quality of expression of written reports, scientific papers, or books. It is imperative to know that the mastery of academic writing methods is not just a pragmatic skill, but also one that can lend integrity and authority to the scholarly research and communication process and establish a researcher's or a team of researchers' obligation to the principle of intellectual honesty in scholarship.

5.4.2   Citations

A *citation* is a quotation from or reference to a source such as a book, dissertation, scientific article, or an author or a team of authors, in particular, in a scholarly work. These sources are cited in the main text of the writing in various ways. At the same time, the referenced publications' complete bibliographical details are listed in the reference section, quoting the *name and initials of the author*, *publication year*, the *publication title, publisher,* and *volume and page numbers* (where applicable).

5.4.3   Bibliography and References

A *bibliography* is a list of all relevant information sources that were studied, though not cited, and might not have been employed in the scholarly works [2]. It is usually printed as an appendix (on a separate page) at the end of an essay, titled '*bibliography*'. *References* provide (a book or an article) with citations of all sources of information that are cited in scholarly works. These citations of the sources of information are carried out by in-text citing in the body of the main text and then listing (the full citations) at the end of the main text under the '*references section*'. Unlike bibliography, references provide a reference list for all in-text citations in the main text that must match an entry in the list [10].

There are two basic standard referencing styles for in-text citations and reference listing in research projects, journal articles, or book publications. These are the *Harvard* (author-date) and *Vancouver* (author-number) styles. Numerous professional publishers typically have their own in-house style, which introduces individual variations from these general conventions.

5.4.3.1 Harvard Style

The *Harvard style* of referencing comprises the citations of the information sources by the in-text citing of the *author's name and publication year* in the main text and alphabetically listing the full citations at the end of the main text under the references section.

5.4.3.2 Vancouver Style

Unlike the Harvard style, the *Vancouver style* involves the assigning of *a number series* in parenthesis for in-text citations of the individual sources of information as cited and the listing of the full citation in numerical order at the end of the main text under the references section. The initial number assigned to a reference is reused each time the assigned reference is cited in the body of the main text, notwithstanding its subsequent position.

5.4.4 Writing Bibliography, Citing and Listing References

Referencing principles using either the *Harvard, Vancouver,* or *other professional in-house* styles involves different referencing for books, journal articles, conference papers, reports, virtual learning environments, the Internet, and so on [11]. For *books*—the *surname* and *initials of the author, publication year*, the *book's title* (in italics or underlined), and the *edition* (if appropriate), *publisher, town,* and *country of publication* are listed, i.e.:

> Ajimotokan, H.A. (2016). *Engineering workshop technology*, 4th Edition. Haajims Publications, Ilorin, Nigeria.

As listed, this book was published in the year 2016 and it is the fourth edition. The book edition is listed as a result of the amount of substantial upgrading and incorporation of new writing contained in a new edition. At the same time, a reprint is just the production of more copies of the original publication. Thus, only new editions require listing during referencing. For *chapters in books*—the *surname* and *initials of the author, publication year*, the *chapter's title*, the *book's title* (in italics or underlined), *editor's initials* and *surname, publisher, town,* and *country of publication* are listed, i.e.:

Ajimotokan, H.A. and Abdulkarim S. (2016). Introduction to workshop practice, Chapter 1, In: *Engineering workshop technology*, H.A. Ajimotokan (Ed.), 4th Edition. Haajims Publications, Ilorin, Nigeria.

This chapter in a book is edited by *Ajimotokan,* and as such, '*(Eds)*' is added after the editor's name. After the '*In*', the convention is to list the book title, followed by the initials and surname of the editor, or vice versa. Haajims Publications is the name of the publisher, and Ilorin and Nigeria are the town and country of publication, respectively. For *journal articles,* the *surname* and *initials of the author, publication year*, the *article's title* (often in inverted commas), the *journal's title* (typically in italics or underlined, though often not), the *journal's volume number, issue,* and page numbers are listed, i.e.:

Ajimotokan, H.A. (2016). 'System dynamics approach for managing magnitude and risk factors of injuries in manufacturing', *Ergonomics,* **57**(13), 45–53.

The volume '*57*' is typically in bold type, the issue number '*13*' is often listed in brackets following the journal's volume number (i.e., '*57(13)*'), and the journal outlet published this article in 2016 in the journal—*Ergonomics.* The article page numbers '*45–53*' should always be listed. To *reference web-based journal articles* and *other materials*, record the *electronic journal article* as well as list the full bibliographic details as mentioned above. Additionally, indicate that the source was the Internet, i.e. the *surname* and *initials of the author, publication year* (if available), the *article's title*, [online] in square brackets (which is optional), journal's information (journal's title in italicised or underlined), available at or retrieved from the *universal resource locator* (URL) of the website, and the *date accessed* (in brackets) are listed. The URL, a distinctive server address, is the identifier for the document. For instance,

Ajimotokan, H.A. (2016). 'System dynamics approach for managing magnitude and risk factors of injuries', *Ergonomics,* **57**(13). Available at: https://en.system.dymanics.approach. for.managing.magnitude.&.riskfactors.of.injuries/57(13) (accessed December 3, 2016).

Ajimotokan, H.A. (2015). MEE 836: Research techniques, *Unilorin Open Courseware,* Dept. of Mechanical Engineering, University of Ilorin, Ilorin, Nigeria. Available at: www.ocw.uni lorin.edu (accessed December 3, 2016).

Research. Forms of research, In: *Research,* Wikipedia the free Encyclopedia. Available at: https://en.wikipedia.org/wiki/Research (accessed December 3, 2016).

The practise of listing references may vary slightly, but still, the imperative is that the important information should at all times be stated such that any researcher is capable of retrieving the article from the database [11].

5.4.5   Citation Management

There are varieties of bibliographic software packages used for bibliography citation management. This bibliographic software collects, stores and organises citations from books, scientific articles, websites and other sources, and automatically converts citations into properly formatted bibliographies, in any referencing style. The bibliographic software saves the author or a team of authors the rigour of citation management, in-text citing and listing of references. Considering the referencing style, and the vast number of other professional in-house variations, some bibliographic software packages allow the storage of PDFs for research papers. Most programs will enable the export or transfer to other programs. Mendeley, EndNote, RefWorks, ProCite, Reference Manager, Zotero and BibDesk, among others are examples of bibliographic software used for creating, recording, editing and electronically storing bibliography citations. Once this software builds a citation library, the user can utilise it time and time again to generate bibliographies for both in-text citing and reference listing, in scholarly articles and essays. A number of these are freely obtainable, like Mendeley, Zotero and BibDesk among others. This software is designed to allow users to create, edit, store and organise citations, including file attachments, in-text citings and listing references from its library. They are a valuable and convenient tool for in-text citing and listing references. The step-by-step procedures for using Mendeley are discussed in the subsequent sub-heading.

5.4.6 Using the Mendeley: Step-By-Step Procedures

Mendeley, a desktop and web program, is utilised to organise and share research articles, discover research data and collaborate online [12]. Mendeley integrates the Mendeley desktop—an application for reference management, with Mendeley Web—an online researchers' social network and a PDF. Using Mendeley demands users to create and store all the necessary data for a citation on its server. It also allows the storage of PDF documents' copies at users' discretion and provides an opportunity to improve the ways references and bibliography are created, organised and generated. Mendeley offers users, upon registration, a 2 GB of free web storage space, which can be upgraded at a cost. To create a Mendeley library,

(i)   Download and install 'Mendeley Software';
(ii) Create an account using an email address;
(iii) Click on the 'Plugin Tab' in Microsoft Word; and
(iv) Select 'Mendeley', and 'Show Mendeley Toolbar', and then click.

Getting started, here are some definitions of terms:

Mendeley Web: This is the website of Mendeley where the user accesses the web version of the library, carries out profile editing, and searches and organises papers, groups, or people. Also, the user might access the social feats of Mendeley from the Mendeley web.

Sync: This is the process of synchronising Mendeley data across devices. The sync makes it easy for users to have every piece of research waiting for them on their PC at home and work, or mobile device—exactly they left it.

Web Importer: This is the browser bookmarklet that permits a quick import of documents from anywhere on the web.

Citation Plugin: This is a plugin installed to permit the creation and formatting of citations and bibliographies according to the chosen style. The Mendeley citation plugin permits in-text citing seamlessly without leaving the word processor, saving time during paper writing by automatically creating the bibliography with a few clicks. This plugin is compatible with MS Word, LibreOffice, and BibTeX documents, allowing the insertion of citations and bibliographies into them.

5.4.6.1 Mendeley Desktop Interface

The *Mendeley desktop* is the downloaded part of the Mendeley software, installed onto a PC. Figure 5.2 depicts the library view of the Mendeley desktop interface of a PC, where 1—the *Add Files Menu*, utilised for adding fresh entries to the library of the Mendeley; 2—the *Folders Menu*, which utilises the '*Add folder*' button for creating a new folder to organise the library of the Mendeley. The '*Add folder*' creates a new folder within whatsoever is presently viewed—if utilised on All Documents, it creates a fresh top-level folder. If utilised on an existing folder, it creates a nested folder in the folder as the parent; 3—the *Sync* is utilised for forcing the Mendeley to perform sync pushing any changes made to the library unto the cloud for storage and available to other devices. The sync should be performed frequently to make sure that all current modifications to the Mendeley library are saved to the cloud; 4—the *Search* is used to search the library of the Mendeley.

Fig. 5.2 The Mendeley desktop interface of the library view. *Source* (Mendeley [13])

The user should be aware that the search function of the Mendeley desktop is context-specific, i.e. searching whilst looking at a specific folder would simply search within that folder. The users of the Mendeley desktop should ensure that they select the '*All documents*' for the search of the whole library. Also, the search function of the Mendeley desktop would return search results for the PDF documents' text(s) within the library, as well as the detailed contents of the document; 5—the *Mendeley (Discovery)* is utilised to find new references in numerous ways. From within the Mendeley desktop, a '*literature search*' can be conducted searching within the crowd-sourced catalogue of Mendeley. Otherwise, the '*Mendeley suggest*' may be utilised to receive adapted recommendations for a specific user on the basis of the user's specific areas of research interests and the Mendeley library's contents; 6—the *My Library* is utilised to view the whole Mendeley library's contents by selecting 'All documents'.

Also, Mendeley provides numerous means for filtering the library. User can also list any created folder(s) under the 'My Library' section; 7—the *Groups* lists any group or groups joined or created under this heading. The icon that appears before the name of the group specifies the group type; 8—the *Main Panel* is utilised to display the contents of the selected view. All entries with an attached PDF are implied by icons that can be double-clicked to open it in the PDF reader; 9—the *Details Panel* is the Mendeley desktop's panel situated furthest to the right, depicting the details of the document type, along with their contents for whichever currently selected library entry. The '*Details panel*' can be utilised to modify an entry's details; and 10—the *Filter Panel* offers numerous alternatives to allow a quick filter of the current view (see Fig. 5.2).

NB: Similar to search, the quick filter of the current view is context-specific, i.e. the panel would simply exhibit the filter options regarding the current view. For instance, when a specific folder is viewed, the user can only view the authors of entries within that folder that appears as options for the filtering. The users of the Mendeley desktop should make sure that they select the '*All documents*' view for the filter of the whole library.

5.4.6.2 Adding Documents

To add a paper entry to the Mendeley library, drag and drop a PDF file into the Mendeley desktop window. Automatically, Mendeley extracts the details from the PDF document and creates a library entry. Similarly, if a folder containing multiple PDF files is dragged and dropped into the window of the Mendeley desktop, Mendeley automatically jockeys the PDF contents it finds in the folder into the library entries. To add a specific file entry to the Mendeley library [13]:

(i)   Click on the '*File tab*' in the Mendeley desktop;
(ii) Select the '*Add files*' from the '*Submenu*' and then click;
(iii) Select the specific paper from the available options; and

Fig. 5.3 The Mendeley
desktop interface of the file
entry view. *Source*
(Mendeley [13])

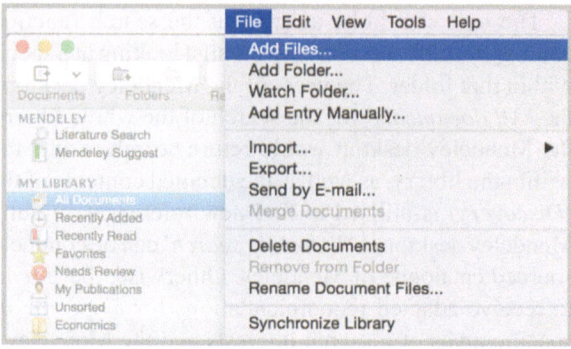

(iv)   Double-click or click '*Open*' to create a library entry for the file in the Mendeley
       library (see Fig. 5.3).

   Similarly, to add entries of a specific folder to Mendeley library,

(i)   Click on the '*File tab*' in the Mendeley desktop;
(ii)   Select the '*Add folder*' from the '*Submenu*' and then click;
(iii) Select the specific folder from the available options; and
(iv)   Click '*Ok*' to create library entries for the folder in order to add any PDF files
       it contains to the Mendeley library.

5.4.6.3   Adding Entry Manually

A user can make a manual entry by clicking on the '*File tab*', '*Add entry manually*',
selecting the appropriate document type, and filling in the fields in the Mendeley
library to create a library entry with the document's reference details. The manual
entry can be utilised to generate library entries for items that do not exist in PDF
form, e.g. books, articles, and dissertations. To add the entry manually,

(i)   Click on the '*File tab*' in the Mendeley desktop;
(ii)   Select the '*Add entry manually*', and then click;
(iii) Selecting the appropriate document type in the '*New document*' and manually
       fill its fields as appropriate; and
(iv)   Save the '*New document*'.

   Once the '*New document*' is saved, the new library entry appears in the Mendeley
library.

   Tip: The document identifier such as DOI, PMID and ArXiv ID can be utilised to
search for the details of a reference. To utilise any of these identifiers, copy and paste
it into the appropriate search field of the '*Discovery*', and click on the magnifying
glass to search the Mendeley catalogue for the item. The '*Discovery*' returns the
details of the item offered by other Mendeley users or from the convenient identifier
service if the reference is entirely new to Mendeley.

5.4.6.4 Adding Entry Electronically

The electronic or automatic input of bibliography citations into a Mendeley library is, in particular, accessible if the article is seen from an online journal databases like ScienceDirect, Scopus, Compendex (Engineering Village), and Google Scholar. To add an entry to the Mendeley library electronically,

(i)   Search the database;
(ii) Select the most relevant article based on the title;
(iii) Click on the '*Export*' or '*Cite article*' from the '*Submenu*';
(iv) Select the '*Mendeley*'; and
(v)   Click on the '*Export*' to import citations directly into the Mendeley library or Mendeley to download a '*Mendeley desktop file*', and then copy and paste the downloaded '*Mendeley desktop file*' into the Mendeley library, creating a library entry for the '*Mendeley desktop file*'.

5.4.6.5 Importing from Other Reference Managers

A user can import library entries from other reference managers such as RefWorks, EndNote, Zotero, ProCite, and Reference Manager, among others. To import from other reference managers,

(i)   Use the '*Export*' option within the other software to extract the citations into RIS, BibTeX or EndNote XML file format;
(ii) Click on the '*File tab*' in Mendeley Desktop;
(iii) Select '*Import*', and then click; and
(iv) Select the exported file to create the library entries of its contents for the Mendeley library.

5.4.6.6 Reviewing the Entries

Whenever a new library entry is entered into the Mendeley library, it is highly recommended to review its details for accuracy. If there is a need to amend the details of a selected library entry, click on the library entry in the '*Main panel*' of Mendeley's Desktop, and then the '*Details panel*' to apply the required changes. The ability to cite and list references accurately is dependent on the accuracy of these library entries, thus, it is recommended to take time to thoroughly review new additions.

5.5 Chapter Summary

Project report writing and its presentation and publication is the concluding requirement of any research and communication process, which researchers, stakeholders, etc. typically interpret as a failure to undertake the research project and its significant findings were, maybe, basically flawed, unsuccessful and irrelevant. The presentation of research outcomes, in particular, the commercial conclusions oriented, and their publication in the reputable publishing outlets of international standard, validate them. Also, it offers objective assistance for research and technology development of databases, strategies or devices, and their commercialisation; and the authorisation that such findings can be suitable additions to the archival literature, engineering standards or both.

Patent writing, a process that requires the technical skills of an expert in the profession, is often written and submitted employing a patent attorney ahead of any conference presentation or journal paper submission. So, a researcher or a team of researchers can no longer patent patentable ideas, innovations, inventions or research outcomes if the idea has been presented in an open forum. Any written report, conference or journal presentation and publication of a research project and its findings follow a similar standard format. The format follows the hourglass structural outline where the introductory and concluding sections must elucidate the project context in the broader discipline. At the same time, the mid-section should focus on the details of the research methodology and results during the presentation.

The use of copied information by representing them in personal writings without the acknowledgement of the source, i.e. plagiarism may put researchers in both ethical and professional problems.

References

1. Thiel, D. V. (2014). *Research methods for engineers*. Cambridge University Press.
2. Lues, L., & Lategan, L. O. K. (2006). *RE: Search ABC* (1st ed.). Sun Press.
3. Booth, W. C., Colomb, G. C., & Williams, J. M. (2008). *The craft of research*. University of Chicago Press.
4. Snieder, R., & Lamer, K. (2009). *The art of being a scientist: A guide for graduate students and their mentors*. University Printing House, University of Cambridge.
5. Alley, M. (2003). *The craft of scientific presentations: Critical steps to succeed and critical errors to avoid*. Springer-Verlag.
6. Hofmann, A. H. (2009). *Scientific writing and communications: Papers, proposals and presentations*. Oxford University Press.
7. University of Oxford. (2019). Plagiarism. Retrieved from https://www.ox.ac.uk/students/academic/guidance/skills/plagiarism?wssl=1
8. Neville, C. (2007). *The complete guide to referencing and avoiding plagiarism*. Open University Press.
9. Walliman, N. (2011). *Research methods: The basics*. Routledge—Taylor and Francis Group.
10. Woods, G. (2002). *Research papers for dummies*. Hungry Minds.
11. Bell, J. (2010). *Doing your research project: A guide for first-time researchers in education and social science* (5th ed.). Open University Press.
12. Mendeley. (2018). Mendeley. *Wikipedia the Free Encyclopedia*. Retrieved from https://en.wikipedia.org/wiki/Mendeley
13. Mendeley. (2018). *Getting started with Mendeley desktop*. Retrieved from https://www.mendeley.com/guides/desktop